Ore stringers or veinlets with greisen-type alteration rims. From "Mineralogische Geographie der Chursächsischen Lande", J. F. W. CHARPENTIER, Leipzig, 1778.

Introduction

A "Glossary of Mining Geology" needs two types of Introductions: First, its "raison d'être" has to be pointed out; second its scope and limitations need to be stated.

A "raison d'être" or justification of its existence is probably easily given. After all, a Glossary published in four languages just joins into the general effort of new International Roof Societies and Journals, to join together the many and diverse groups of the same professional interest. Our Glossary should help to promote better understanding and communication in the field of mineral exploration, mining geology and mineral research.

By the same token, an effort was made to keep in pace with the present evolution of thought in Ore Genesis, not by bending definitions into the opposite extreme direction, but by neutralizing some and leaving open others, where a satisfactory solution to the problem of genesis has not been found yet. –

The second introduction, referring to the scope and limitations, is best expressed with quotations from the famous lectures of ORTEGA Y GASSET on the "Misery and Splendor of Translation" (miseria y esplendor de la traducción). In the following paragraph ORTEGA points out the difficulty of finding exact synonyms in another language: hardly any term is entirely congruent with its translation simply because it has normally had a different history.

"Every language, when compared to other languages, has its own linguistic style which HUMBOLDT calls its "inner form". It is utopian therefore to believe that two words belonging to two different languages and being offered as translations in a dictionary actually have the same meaning. Since languages develop in different countries and under different living conditions and éxperiences the incongruences are quite natural. For example, it would be quite wrong to think that what the Spaniard calls a bosque means exactly the same thing that a German calls a Wald. Though the dictionary tells us that Wald means bosque. And here would be a perfect opportunity to insert a Bravura-Aria describing the difference between the German Wald in constrast to the Spanish bosque I shall forget this song, but must insist on its conclusion, i. e. upon the clear understanding in the colossal difference existing between both actualities". (OBRAS V, p. 436).

In a later paragraph ORTEGA states that man insists in attempting to do the impossible, and that one of these impossible things is a perfect translation:

"History lets us realize mankinds incessant and inexhaustable ability to invent projects which cannot be realized. In his efforts to realize them he accomplishes much, creats infinite realities which so called nature finds impossible to create. The only thing that man is never able to accomplish, is that which he plans to undertake, which must be said in his favour. This marriage of reality with the impossible creates the only thing he is able to add to the universe.

That ist why it is so important to underline that all–to be sure all which is worth the effort, all which is human–is difficult, very difficult, so much so that it is impossible. As you see, to declare its impossibility is not in itself an objection against the possible splendor of translating. On the contrary, this property lends it the most sublime affiliation and lets us anticipate that there is sense to it." (OBRAS V, p. 439).

The Spanish philosopher then continues to inform us about the relativity of language. He says that even in physics man is often not able to express what he really means. Between the lines one can read how well aware he was of the subconscious sources and roots of many terms even in the sciences:

"We assert that man, when he begins to speak, does so while he believes that he may say

V

what he thinks. But that is precisely what is so misleading. Speach does not accomplish that much. It expresses more or less a portion of what we think and places the remainder behind an insurmountable barrier. It easily suffices for mathematical concepts and proofs whereas in physics it begins to be ambiguous and insufficient. And it increases in its inexactness, obtuseness and tendency to confusion in direct proportion to its use in discussions concerning mankinds more "actual" topics" (OBRAS V, p. 442).

By these quotations I want to emphasize how well aware of the relativity of language and of the difficulties of translation I am. Nevertheless, international communication requires translation and this was one main reason for setting up this Glossary of Mining Geology. But the relativity of a terminology is not only one of place, i. e., of language and culture, but also one of time. Definitions change with the development of sciences and the ones offered in our Glossary are only meant to be working hypotheses. I hope, naturally, that most of them are at the "height of our time".

Consequently, the definitions are not meant to be stiff dogmas. We all have the innate tendency to snuggle up to the available warm security of classifications and definitions. To compartmentalize according to readily available textbook-definitions is less threatening to the intellectual equilibrium than the pursuit of the challenge into the double roots of definitions: the nature of the object outside of us—rocks, minerals and ore deposits—and the nature of the subconscious source of our more or less intuitive interpretations. The definitions given in this Glossary should not chloroform the phenomena and hide or freeze the problems existing in very many instances. It is hoped that it will have the opposite effect: a mobilization of the discussion, an information on the existence of more than one interpretation.

For all these reasons the author and his friends who so generously helped to build this Glossary to the present level, wish to request open and critical comments and additions for a second edition.

G. C. AMSTUTZ
Heidelberg
March 1970

Introduccion

Un Glosario de Geología de Minas necesita dos tipos de introducción: en primer lugar se debe especificar su razón de ser, y en segundo lugar se necesita establecer su objeto y limitaciones.

Una razón de ser o justificación de su existencia se puede dar con facilidad. Después de todo un glosario publicado en cuatro idiomas recopila los esfuerzos generales de nuevas Sociedades y Revistas Internacionales, con objeto de reunir los muchos y diversos grupos que tengan el mismo interés profesional. Nuestro glosario ayuda a promover un mejor entendimiento y comunicación en el campo de la exploración minera, geología de minas, e investigación mineral.

A la vez se ha realizado un esfuerzo para estar a la altura de la presente evolución de conceptos en la génesis mineral, no fijando definiciones en aquellos casos que sean opuestos, sino anulando algunos y dejando otros abiertos, cuando no se haya encontrado aún una solución satisfactoria al problema de la génesis.

La segunda introducción, referente al objeto y limitaciones, se expresa mejor mediante citas de las famosas lectura de la obra de ORTEGA Y GASSET "miseria y esplendor de la traducción". En el párrafo que se cita a continuación ORTEGA señala la dificultad de encontrar sinónimos exactos en otro idioma: apenas ningún término es totalmente congruente con su traducción, simplemente a causa de que ha tenido una historia distinta.

". . . es el caso que cada lengua comparada con otra tiene también su estilo lingüístico, lo que Humboldt llamaba su "forma interna". Por tanto, es utópico creer que dos vocablos pertenecientes a dos idiomas y que el diccionario nos da como traducción el uno del otro, se refieren exactamente a los mismos objetos. Formadas las lenguas en paisajes diferentes y en vista de experiencias distintas, es natural su incongruencia. Es falso, por ejemplo, suponer que el español llama bosque a lo mismo que el alemán llama *Wald*, y, sin embargo, el diccionario nos dice que *Wald* significa bosque. Si hubiera humor para ello sería excelente ocasión para intercalar un "aria de bravura" describiendo el bosque de Alemania en contraposición al bosque español. Hago gracia a ustedes de la canción, pero reclamo su resultado: la clara intuición de la enorme diferencia que entre ambas realidades existe." (OBRAS V, p. 436).

En un párrafo posterior Ortega establece que el hombre insiste en intentar hacer lo que es imposible, y una de las cosas que son imposibles es una traduccion perfecta:

"La historia universal nos hace ver la incesante e inagotable capacidad del hombre para inventar proyectos irrealizables. En el esfuerzo para realizarlos logra muchas cosas, crea innumerables realidades que la llamada naturaleza es incapaz de producir por sí misma. Lo único que no logra nunca el hombre es, precisamente, lo que se propone—sea dicho en su honor. Esta nupcia de la realidad con el íncubo de lo imposible proporciona al universo los únicos aumentos de que es susceptible. Por eso importa mucho subrayar que todo—se entiende todo lo que merece la pena, todo lo que es de verdad humano—es difícil, muy difícil; tanto, que es imposible.

Como ustedes ven, no es una objeción contra el posible esplendor de la faena traductora declarar su imposibilidad. Al contrario, este carácter le presta la más sublime filiación y nos hace entrever que tiene sentido." (OBRAS V, p. 439).

El filósofo español continua entonces informandonos acerca de la relatividad del lenguaje. El dice que incluso un físico a menudo no está capacitado para expresar lo que realmente piensa. Entre lineas uno puede leer qué interés le ha prestado a las raices y procedencias subconscientes de muchos términos parecidos que hay en las ciencias:

"Digamos, pues, que el hombre, cuando se pone a hablar lo hace *porque* cree que va a poder decir lo que piensa. Pues bien; esto es ilusorio. El lenguaje no da para tanto. Dice, poco más o menos, una parte de lo que pensamos y pone una valla infranqueable a la transfusión del resto. Sirve bastante bien para enunciaciones y pruebas matemáticas: ya el hablar de física empieza a ser equivoco o insuficiente. Pero conforme la conversación se ocupa de temas más importantes que ésos, más humanos, más "reales", va aumentando su imprecisión, su torpeza y su confusionismo." (OBRAS V, p. 442).

Con estas citas yo deseo subrayar la atención que he prestado a la relatividad del lenguaje y a las dificultades de traducción. Sin embargo las comunicaciones internacionales requieren ser traducidas, y esta ha sido una razón importante para crear este Glosario de Geologia de Minas. Pero la relatividad de una terminología depende no solo del lugar, como le ocurre al lenguaje y la cultura, sino que tambien depende del tiempo. Las definiciones cambian con el desarrollo de las ciencias y las ofrecidas en este glosario se han pensado solamente como hipótesis de trabajo. Espero, naturalmente, que la mayoria de ellas esten a la "altura de nuestro tiempo".

Por consiguiente las definiciones no pretenden ser dogmáticas. Todos tenemos la tendencia innata a acomodarnos a una cierta seguridad en las clasificaciones y definiciones útiles. Subdividir de acuerdo con una forma comoda las definiciones de un texto es menos arriesgado para el equilibrio intelectual que el buscar la razón del doble origen de una definición: la naturaleza del objeto que nos rodea −rocas, minerales, menas−, y la naturaleza de las fuentes subconscientes de nuestras interpretaciones más o menos intuitivas. Las definiciones dadas en este glosario no debieran apartarnos del fenómeno ni de su importancia, ni hacernos olvidar los problemas existentes en muchos ejemplos. Es de esperar que tenga un efecto contrario, es decir, sea una tendencia a la discusión, una información de la existencia de más de una interpretación.

Por todas estas razones el autor y sus compañeros, que tan generosamente han colaborado en la confección de este Glosario al nivel actual, desean sugerir se hagan adendas y comentarios abiertos y críticos para una segunda edición.

G. C. AMSTUTZ
Heidelberg
Marzo 1970

Avant-Propos

Un «Glossaire de Géologie Minière» demande deux sortes d'avant-propos: en premier lieu dire sa raison d'être, en second, exprimer son but et ses limitations.

Sans doute est-il facile de justifier son existence ou de donner sa raison d'être. Après tout, un glossaire en quatre langues s'intègre parfaitement dans l'effort général de création de sociétés et de journaux internationaux tendant à rassembler les nombreux et divers groupes de personnes, ayant des intérêts professionnels communs. Notre glossaire devrait aider la compréhension et la communication des idées, dans les domaines de l'exploration et de la recherche minérale et de la géologie minière.

Par conséquent nous nous sommes efforcés de suivre de près l'évolution actuelle de la pensée en métallogénie, non pas en modifiant les définitions dans un sens diamétralement opposé, mais en neutralisant certaines et laissant d'autres ouvertes à la discussion, quand une solution satisfaisante n'a pas encore été trouvée à ce jour au problème de la genèse.

Des citations extraites des fameuses conférences d'ORTEGA Y GASSET sur la «Misère et Splendeur de la Traduction», constituent au mieux le deuxième avant-propos traitant du but et des limitations. Dans le paragraphe suivant, ORTEGA souligne la difficulté de trouver des synonymes exacts dans une autre langue: peu de termes sont entièrement concordants (congruents) avec leur traduction, tout simplement parce qu'ils ont des histoires différentes.

«. . . Chaque langue comparée à une autre, possède également son style linguistique, que HUMBOLDT appellait sa «forme interne». Par conséquent, il est utopique de croire que deux vocables appartenant à deux langues, et que le dictionnaire nous donne comme traductions l'un de l'autre, se réfèrent exactement aux mêmes objets. Comme les langues ont été formées dans des paysages différents et tenant compte d'expériences distinctes, une telle inadiquation (incongruence) est naturelle. Il est faux, par exemple, de supposer que l'espagnol appelle bosque (forêt) la même chose que l'allemand appelle wald, et cependant le dictionnaire nous dit que wald signifie bosque. Si je m'y sentais disposé, ce serait là une excellente occasion d'intercaler ici un «air de bravoure» décrivant la forêt allemande par opposition à la forêt espagnole. Je vous fais grâce de la chanson, mais je revendique son résultat, c'est-à-dire la claire intuition de l'énorme différence qui existe entre ces deux réalités.» (OBRAS V, p. 436).

Dans un paragraphe ultérieur, ORTEGA dit que l'homme insiste dans ses tentatives à faire l'impossible et que l'une de ces impossibilités est une traduction parfaite.

«. . . L'histoire universelle nous montre l'incessante et l'inépuisable capacité de l'homme à inventer des projets irréalisables. Dans son effort à les réaliser, il réussit beaucoup, créé d'innombrables réalités que la dite nature est incapable de réaliser par elle-même. La seule chose que l'homme ne réussit jamais, c'est précisément ce qu'il se propose − soit dit en son honneur. Ce mariage de la réalité avec le démon de l'impossible procure à l'univers les uniques augmentations dont il est susceptible. A cause de cela il importe beaucoup de souligner que tout − c'est-à-dire tout ce qui en vaut la peine, tout ce qui est véritablement humain − est difficile, très difficile, à tel point qu'il est impossible. Comme vous voyez, déclarer que le travail du traducteur est impossible n'est pas une objection contre son éventuel éclat. Au contraire ce caractère lui confère une filiation sublime et nous fait entrevoir qu'il a une signification.» (OBRAS V, p. 439).

Ensuite le philosophe expose la relativité du langage. Il dit que même en physique, l'homme est souvent incapable d'exposer sa pensée. On peut lire entre les lignes combien il était conscient des sources et des racines subconscientes de nombreux termes, même scientifiques.

«. . . Disons donc que l'homme, quand il se met à parler, le fait parce qu'il croit qu'il va pouvoir dire ce qu'il pense. Soit, mais cela est illusoire. Le langage ne va pas si loin. Il rend, plus ou moins, une partie de ce que nous pensons et oppose une barrière infranchissable au passage du reste. Il sert assez bien aux énoncés et démonstrations mathématiques: déjà le langage de la physique commence à être ambigu ou insuffisant. Mais plus la conversation se tourne vers des thèmes plus importants que ceux-ci, plus humains, plus «réels», plus augmentent son imprécision, sa gaucherie et sa tendance aux confusions.» (OBRAS V, p. 442).

Par ces citations, je voudrais faire ressortir combien je suis conscient de la relativité de la langue et de la difficulté des traductions. Néanmoins, la communication internationale exige traduction et voici l'une des raisons d'être principales de ce Glossaire de Géologie Minière. Mais la relativité de la terminologie est non seulement liée à l'espace, c'est-à-dire à la langue et à la culture, mais aussi au temps. Les définitions changent avec le développement des sciences et celles proposées dans ce glossaire sont seulement destinées à être des hypothèses de travail. J'espère naturellement que la plupart d'entre elles sont «à jour».

Par conséquent, les définitions ne doivent pas être prises comme des dogmes rigides. Nous avons tous la tendance innée à nous nicher dans la confortable sécurité des classifications et définitions. Classer, en fonction de définitions déjà données par des traités, menace moins l'équilibre intellectuel que pousser le défi jusqu'à rechercher les racines doubles des définitions; la nature intrinsèque de l'objet (roche, minéral et gisement) et la nature de la source subconsciente de nos interprétations plus ou moins intuitives. Les définitions données dans ce glossaire ne devraient pas «chloroformer» les phénomènes et cacher, ou du moins figer, les problèmes qui se posent souvent. Nous espérons que l'effet contraire se réalisera: une mobilisation de la discussion, une information sur la pluralité des interprétations.

Pour toutes ces raisons, l'auteur et ses amis, qui l'ont si généreusement aidé à porter ce glossaire à son niveau actuel, seraient heureux de recevoir des commentaires francs et critiques et des additions en vue de la deuxième édition.

G. C. Amstutz
Heidelberg
Mars 1970

Vorwort

Ein „Wörterbuch der Montangeologie" benötigt zweierlei Einleitungen: Einmal eine Darlegung seiner „raison d'être" und zweitens einen Hinweis auf seine Ziele und Grenzen.

Seine Daseinsberechtigung dürfte kaum bezweifelt werden in einer Zeit, wo die internationalen Verbindungen so sehr intensiviert und die Notwendigkeiten fachlicher Kommunikation alltäglich geworden sind. Das vorliegende Glossary möchte hier eine Lücke füllen.

Gleichzeitig wurde versucht, Schritt zu halten mit der gegenwärtigen Entwicklung auf dem Gebiet der Lagerstättengenese, und zwar nicht durch Übertreibungen in entgegengesetzten Richtungen. Einzelne Ausdrücke wurden neutralisiert und andere einfach offen gelassen, und zwar stets dann, wenn sich eine vernünftige genetische Entscheidung noch nicht abzeichnet.

Nun ist aber das Übersetzen an sich ein äußerst problematisches, ja sogar fragwürdiges Unterfangen. Um beim Leser und Benutzer keine Zweifel aufkommen zu lassen über die Grenzen und Mängel der Übersetzerarbeit an sich, möchte ich das Büchlein nicht hinausgehen lassen ohne ausdrückliche Zitate aus dem klassischen Vortrag von ORTEGA Y GASSET über „Glanz und Elend der Übersetzung" (weitgehend zitiert nach der deutschen Übersetzung der „Gesammelten Werke in vier Bänden" der Deutschen Verlags-Anstalt Stuttgart). – ORTEGA macht uns hier klar, daß kaum ein Ausdruck einen congruenten Gegenwert in einer anderen Sprache findet, ganz einfach deshalb, weil die Sprachentwicklung kulturtypisch ist, also in jedem Lande anders.

„Es verhält sich so, daß jede Sprache im Vergleich mit einer anderen ebenso ihren besonderen sprachlichen Stil besitzt, das, was Humboldt ihre ‚innere Form' nannte. Aus diesem Grund ist es utopisch, zu glauben, daß zwei Wörter, die zwei verschiedenen Sprachen angehören und die uns das Wörterbuch als ihre wechselseitige Übersetzung darbietet, genau die gleichen Dinge bedeuten. Da die Sprachen in verschiedenen Landschaften und unter dem Einfluß verschiedener Lebensumstände und Erfahrungen gebildet wurden, ist ihre Inkongruenz ganz natürlich. So ist es z. B. falsch, anzunehmen, daß das, was der Spanier b o s q u e nennt, das gleiche sei, was der Deutsche Wald nennt, und doch sagt uns das Wörterbuch, daß Wald b o s q u e bedeutet. Wenn ich Lust hätte, wäre hier eine vortreffliche Gelegenheit, eine Bravour-Arie einzulegen, die den deutschen Wald im Gegensatz zum spanischen Wald beschreibt. Ich schenke Ihnen das Lied, bestehe aber auf seinem Schluß, d. h. auf der klaren Einsicht in den enormen Unterschied, der zwischen diesen beiden Wirklichkeiten besteht." (OBRAS V, p. 436.)

In einem weiteren Abschnitt macht uns der spanische Philosoph und Schriftsteller klar, daß es eine ureigene menschliche Eigenschaft ist, das Unmögliche zu versuchen, und daß er deshalb stets wieder versucht, perfekte Übersetzungen herzustellen.

„Die Weltgeschichte läßt uns die unaufhörliche und unerschöpfliche Fähigkeit des Menschen erkennen, Projekte zu erfinden, die nicht verwirklicht werden können. In dem Bemühen, sie zu verwirklichen, erreicht er vieles, erschafft er unzählige Realitäten, die die sogenannte Natur unfähig ist, aus sich selbst hervorzubringen. Das einzige, was der Mensch niemals erreicht, ist eben das, was er sich vornimmt – zu seiner Ehre sei es gesagt. Diese Vermählung der Wirklichkeit mit dem Inkubus des Unmöglichen schafft dem Universum die einzigen Erweiterungen, deren es fähig ist. Darum ist es so wichtig, zu unterstreichen, daß alles – selbstverständlich alles, was der Mühe wert ist, alles, was wirklich menschlich ist – schwierig, sehr schwierig ist, so sehr, daß es unmöglich ist.

Wie Sie sehen, spricht die Erklärung ihrer Unmöglichkeit nicht gegen den möglichen

Glanz der Übersetzerarbeit. Ganz im Gegenteil, dieser Charakter verleiht ihr den höchsten Rang und läßt uns ahnen, daß sie von Bedeutung ist." (OBRAS V, p. 439).

ORTEGA informiert uns schließlich über die mannigfaltigen Relativitäten und Beschränktheiten der Sprache und stellt fest, daß sogar der Physiker oft nicht imstande ist, seine Gedanken so auszudrücken, wie er gerne möchte. Zwischen den Zeilen liest man hier deutlich, wie genau er über die unterbewußten Wurzeln und Ursprünge, sogar auch der wissenschaftlichen Terminologie Bescheid wußte.

„Wir sagen also, daß der Mensch, wenn er sich anschickt zu sprechen, es tut, weil er glaubt, das sagen zu können, was er denkt. Nun, gerade das ist trügerisch. Soviel leistet die Sprache nicht. Sie gibt, mehr oder weniger, einen Teil von dem wieder, was wir denken, und setzt der Übermittlung des Restes einen unübersteiglichen Damm entgegen. Sie genügt in ausreichendem Maße für mathematische Begriffe und Beweise, doch beginnt schon die Sprache der Physik zweideutig und ungenügend zu werden. In dem Maße aber, wie die Unterhaltung sich mit wichtigeren, menschlicheren, ‚realeren‘ Themen befaßt, steigert sie ihre Ungenauigkeit, ihre Schwerfälligkeit und ihre Neigung zur Verwirrung." (OBRAS V, p. 442).

Mit diesen Zitaten soll belegt werden, wie sehr wir uns der Relativitäten und Schwierigkeiten der Übersetzungen, sogar auf dem Gebiet der Naturwissenschaften bewußt sind. Trotz dieser Hindernisse schien es angezeigt, im Dienste der internationalen Zusammenarbeit auf dem Gebiete der Lagerstättenforschung und Montangeologie das Glossary zusammenzustellen. Es handelt sich natürlich nicht nur um eine Relativität des Ortes, also der Sprache und Kultur, sondern auch der Zeit. Die Definitionen ändern sich mit der Zeit, d. h. mit der Entwicklung der wissenschaftlichen Vorstellungen. Deshalb sind die Definitionen des Glossary nur als Arbeitshypothesen gedacht. Ich hoffe allerdings, daß die meisten „der Höhe der Zeit" entsprechen.

Auf Grund des Gesagten sollen die Definitionen keine steifen Dogmen repräsentieren. Wir haben natürlich alle die angeborene Tendenz, uns hinter dem sicheren Wall bestehender Klassifikationen und Definitionen zu verschanzen. Dieser Schutz ist für das intellektuelle Gleichgewicht zuträglicher als die offensive Jagd nach neuen, besseren Ideen. Diese Jagd bedingt die Anerkennung der doppelten Wurzel aller bestehenden Definitionen, nämlich der Objekte außerhalb von uns, d. h. der Gesteine, Mineralien und Erze, und der unbewußten, inneren Quelle unserer Interpretationen und Auswahlentscheidungen. Die Definitionen des Glossaries sollen also die Phänomene nicht einbalsamieren und somit die bestehenden Probleme verbergen oder verdrängen. Wir hoffen sogar, daß es gerade den entgegengesetzten Effekt haben wird, nämlich eine Mobilisation der Diskussion, z. B. durch die Information über die Existenz von mehr als einer Interpretationsmöglichkeit.

Auf Grund dieser Gedanken und Gründe möchte der Autor und seine Kollegen, die in sehr großzügiger Weise beim Ausbau des Glossaries beteiligt waren, alle Leser um konstruktive Kritik und um Mitteilung von Lücken bitten, im Hinblick auf die zweite Auflage.

G. C. AMSTTTZ
Heidelberg
März 1970

Actualism or uniformitarianism
(J. Hutton, 1787, C. H. Lyell, 1830): The doctrine that all geological events which took place in the past can be explained by phenomena and processes still active at present. (Sch.)

Actualismo o uniformismo. (J. Hutton, 1787, C. H. Lyell, 1830): Teoría que afirma que todos los acontecimientos geológicos ocurridos en el pasado pueden ser aún explicados por fenómenos y procesos activos en el presente.

Actualisme: Doctrine défendue par J. Hutton (1787) et C. H. Lyell (1830) suivant laquelle toutes les forces et tous les phénomènes géologiques qui prirent place dans le passé se produisent et peuvent être expliqués par des phénomènes et des évènements encore actives aujourd'hui. (On dit aussi «Principe d'actualité» ou «Principe des Causes actuelles».)

Aktualismus: (J. Hutton, 1787, C. H. Lyell, 1830): Die Lehre, nach der alle geologischen Vorgänge, die in der Vergangenheit stattgefunden haben, durch Phänomene und Vorgänge erklärt werden können, die auch gegenwärtig noch bestehen.

Aggregate
(rock): Sand, gravel, broken gravel, crushed stone, or a mixture of these used with cement to make concrete.

Agregado (roca): Arena, grava, arena gruesa, piedra triturada, o una mezcla de ellos usados con cemento para hacerlo concreto.

Agrégat: Sable, gravier, gravillon, roche concassée ou mélange de ceux-ci destinés à faire partie d'un béton par mélange avec un ciment.

Zuschlagstoff (Gestein): Sand, Schotter, Kies, gebrochener Stein oder Gemenge aus diesen, die mit Zement zur Herstellung von Beton oder Mörtel verwendet werden.

Allochthonous:
Tectonic units which originate from outside their present location, or, rock components which did not form *in situ* (term introduced by von Gümbel, 1888) (opp. *autochthonous*).

Alóctono: Unidades tectónicas originadas fuera de su presente ubicación, o componentes de roca que no se formaron *in situ* (término introducido por von Gümbel, 1888) (opuesto: autóctono).

Allochtone: 1. Unité tectonique formée en un endroit nettement différent de son emplacement actuel où elle est parvenue par un transport de grande envergure. 2. Constituant d'une roche qui ne s'est pas formée *in situ* (von Gümbel, 1888) (contraire: autochtone).

Allochthon: Tektonische Einheiten, die von woanders als ihrer gegenwärtigen Fundstätte stammen, oder Gesteinskomponenten, die nicht *in situ* entstanden sind (Ausdruck eingeführt durch von Gümbel, 1888; Gegenteil: autochthon).

Alluvial deposits (see Fig. 14 and *placers*):
Placer material concentrated by stream action.

Yacimientos o Depósitos aluvionales (o aluviales): Material de placeres concentrados por acción de corrientes.

Gîte alluvionnaire: Matériel détritique concentré par l'action des courants.

Alluviale Lagerstätten: Ablagerungen, die durch Wasserströmung angereichert werden (Fluß-Seifen).

1

Alteration:

Designation of minor or major changes in the mineralogical composition of a rock brought about by hypogene or supergene, syngenetic or epigenetic, hydrothermal, deuteric, pneumatolytic or groundwater solutions. Usually distinguished from metamorphism by degree or extent of the changes; corresponding to or grading into epimetamorphism. Alteration may be associated syn- or epigenetically with mineralizations. (Ex.: kaolinization, albitization, chloritization, propylitization, sericitization, etc.)

Alteración: Cambio mas o menos fuerte en la composición mineralógica de una roca, debido a soluciones hipogenica o supergenica, hidrotermales, singenéticas o epigenéticas, deutéricas, neumatolíticas, o a aguas subterráneas. Generalmente se distingue del metamorfismo por el grado o extensión del cambio. Puede identificarse con el epimetamorfismo o pasar progresivamente a el. La alteración puede estar asociada o epigenéticamente a mineralizaciones. (ejemplos: caolinización, albitización, cloritización, propilitización, sericitización, etc.)

Altération: Modification plus ou moins intense de la composition minéralogique d'une roche provoquée par des solutions hypogènes ou supergènes, syngénétiques ou épigénétiques, hydrothermales, deutériques, pneumatolytiques ou des eaux souterraines. L'altération se distingue généralement du métamorphisme par l'intensité et l'étendue des modifications. Elle peut s'identifier ou passer progressivement à l'épimétamorphisme. Elle peut être associée de manière syngénétique ou épigénétique aux minéralisations. (Ex.: albitisation, chloritisation, kaolinisation, propylitisation, séricitisation, etc.)

Umwandlung, Veränderung: Bezeichnung mehr oder weniger intensiver Veränderungen oder Umwandlungen in der mineralogischen Zusammensetzung von Gestein, hervorgerufen durch hypogene oder supergene, deszendente oder aszendente, syngenetische oder epigenetische, hydrothermale, deuterische oder pneumatolytische Lösungen oder durch das Grundwasser. Gewöhnlich unterschieden von Metamorphose durch den Grad oder das Ausmaß der Veränderungen. Kann aber der Epimetamorphose entsprechen oder in sie übergehen. Diese Umwandlungen können syn- oder epigenetisch mit der Vererzung vor sich gehen (z. B. Kaolinisierung, Albitisierung, Chloritisierung, Propylitisierung, Serizitisierung, etc.).

Anaerobic:

Conditions or organisms existing without free oxygen.

Anaeróbia: Condiciones u organismos existentes sin oxígeno libre.

Anaérobie: Conditions ou organismes existant sans oxygène libre.

Anaerob: Zustand oder Organismen ohne freien Sauerstoff.

Anatexis (see Appendix VIId):

A partial, selective fusion of ultrametamorphic rocks (often used as a synonym of *palingenesis* which, however, may be used to designate more complete melting or remelting).

Anatexis: Fusión parcial selectiva de rocas ultrametamórficas (usado a menudo como sinónimo de *palingénesis*; este último término implica sin embargo una fusión o refusión más completa).

2

Anatexie: Fusion partielle et sélective de roches ultramétamorphiques; souvent utilisée comme synonyme de *palingenèse*, ce dernier terme impliquant cependant une fusion ou refusion plus complète.

Anatexis: Eine teilweise selektive Aufschmelzung ultrametamorphen Gesteins. (Oft wird als Synonym *Palingenese* verwandt, die jedoch als noch weitgehendere Aufschmelzung oder Wiederaufschmelzung betrachtet werden kann.)

Apex

(of a vein or lode): Term used in U. S. mining law to denote the outcrop of a vein which reaches the surface, or the highest limit of a vein which does not extend upward all the way to the surface. (MCKINSTRY)

Apice (de un filón o veta): Término usado en las leyes mineras de EE. UU. para denotar la parte del afloramiento de una veta que alcanza la superficie, o la parte más alta de una veta que no se extiende en la superficie.

Tête (d'un filon ou d'un lode): Terme utilisé dans la loi minière américaine pour désigner l'affleurement d'un filon qui atteint la surface ou la limite supérieure d'un filon qui ne s'étend pas jusqu'au jour.

—: Im Bergbaurecht der USA üblicher Ausdruck zur Bezeichnung des Ausbisses eines Ganges oder der obersten Grenze eines Ganges, der nicht ganz an die Oberfläche reicht.

Apomagmatic deposit (see Appendix VIIb):

Formed outside of, but in close proximity to, and in clear connection with a parent igneous body.

Depósito apomagmático: Formado fuera de la roca ignea generatriz, pero muy cerca y en clara conexión con ella.

Gîte apomagmatique: Gîte formé à l'extérieur de la roche ignée qui lui a donné naissance, mais à proximité et en relation évidente avec cette roche.

Apomagmatische Lagerstätte: Außerhalb, aber in großer Nähe und deutlicher Verbindung zu einem ursächlichen Eruptivgestein.

Aquifer:

A water-bearing layer or formation, e. g., a permeable sandstone which when tapped by a well yields a supply of water. (MCKINSTRY)

Acuífero: Capa o formación que surte agua; ejemplo: una arenisca permeable horadada para suministrar agua.

Aquifère: Couche ou formation contenant une nappe, p. ex. un grès perméable qui fournit de l'eau lorsqu'il est traversé par un puits.

Grundwasserleiter: Eine wasserspeichernde Schicht oder Formation, z. B. ein durchlässiger Sandstein, der Wasser abgibt, wenn er durch eine Bohrung angezapft wird.

Ascension theory

(ELIE DE BEAUMONT, 1847): The theory that the matter filling fissure veins was introduced in solution from below. (SCH.)

Teoría ascensional (ELIE DE BEAUMONT, 1847): Admite que las materias que rellenan vetas de fisuras fueron introducidas por soluciones procedentes de abajo.

Ascensionnisme (ELIE DE BEAUMONT, 1847): Théorie admettant que le matériel d'une caisse filonienne a été mis en place par des solutions venant du bas.

Aszendenztheorie (ELIE DE BEAUMONT, 1847): Die Theorie, nach der das gangfüllende Material durch Lösungen aus der Tiefe (aszendent) herantransportiert wurde.

Assay:

Verb: To determine the amount of metal contained in an ore sample. Noun: 1. The act of making such a determination. 2. The result of such a determination. Note: Common difference between assay and analysis: In an analysis all of the chemical constituents are normally determined; in an assay only certain constituents, usually those of commercial interest.

Ensayo: Verbo: determinar la cantidad de metal contenido en una mena. Sustantivo: 1. El acto de hacer tal determinación. 2. El resultado de tal determinación. Nota: Diferencia común entre ensayo y análisis: En un análisis frecuentemente todos los componentes químicos son determinados; en un ensayo solamente ciertos constituyentes, generalmente aquellos de interés comercial.

Dosage (verbe: doser): Détermination de la quantité de métal contenue dans un minerai; le résultat d'une telle détermination est une teneur. N. B. Ce terme diffère de l'analyse quantitative qui comporte normalement le dosage de tous les éléments chimiques. Le dosage d'un minerai n'implique que la détermination des teneurs de certains éléments, généralement ceux d'intérêt commercial.

—: Chemische Probenauswertung in Bergbaubetrieben. Anmerkung: Unterschied zwischen *Assay* und Analyse: In einer Analyse werden gewöhnlich alle chemischen Bestandteile bestimmt; bei dem englischen Ausdruck *Assay* nur einige Bestandteile, und zwar solche, die von wirtschaftlicher Bedeutung sind.

Assimilation

(or magmatic assimilation): The incorporation into a magma, of material originally present in the wall rock. The term does not specify the exact mechanism or results; the "assimilated" material may be present as crystals from the original wall rocks, newly formed crystals including wall rock elements, or as a true solution in the liquid phase of the magma. The resulting rock is called *hybrid*. Term also applicable to ore veins.

Asimilación (o asimilación magmática): La incorporación a un magma de material originalmente presente en la roca de caja. El término no especifica el mecanismo o resultado exacto; el material „asimilado" puede estar presente como cristales de la roca encajante original, cristales recientemente formados incluyendo elementos de la roca de caja, o como verdadera solución en la fase líquida del magma. La roca resultante se llama *híbrida*. El término es aplicable a vetas de menas.

Assimilation (ou assimilation magmatique): Incorporation dans un magma de matériaux de la roche encaissante. Le terme ne précise ni le mécanisme exact, ni les résultats; les matériaux assimilés peuvent être présents sous forme de cristaux de la roche encaissante ou sous forme de cristaux néoformés à partir des éléments de la roche encaissante ou sous forme de véritable solution dans la phase liquide du magma. Le terme d'assimilation peut également s'appliquer aux filons métallifères.

4

Assimilation (oder magmatische Assimilation): Die Einverleibung einer ursprünglich im Nebengestein vorhandenen Materie in ein Magma. Der Ausdruck definiert keinen bestimmten Mechanismus oder genaue Ergebnisse; das assimilierte Material kann aus Mineralkörnern des ursprünglichen Nebengesteins bestehen, aus nachträglich geformten Körnern desselben, oder aus echten Lösungen in der flüssigen Phase des Magmas. Das entstandene Gestein wird *hybrid* genannt. Gebräuchliche Definition auch bei Erzgängen.

Attitude:

Orientation in space, i. e., direction and inclination of a structural plane or line (bed, fault, vein, etc.). Attitude may be expressed in terms of *dip* and *strike*.

Posición: La dirección y el grado de buzamiento de un plano estructural (capa, falla, veta, etc.). La posición puede ser expresada en términos de buzamiento y rumbo (véase allí.)

Disposition: Orientation et valeur du pendage d'un plan structural (couche, faille, filon etc.). La disposition peut être repérée par le *pendage* et la *direction* (voir ces termes).

Lage: Richtung und Winkel des Einfallens einer Strukturfläche (Schicht, Verwerfung, Gang, etc.). Lage kann durch Fallen (= Neigung) und Streichen bestimmt werden (siehe dort).

Autochthonous:

Tectonic units or masses which were formed in, and have never moved appreciably from, the original location. Also: rocks or rock constituents which have formed *in situ* (term created by VON GÜMBEL in 1888) (opp. *allochthonous*).

Autóctono: Unidades o masas tectónicas que fueron formadas y nunca movidas considerablemente de su presente ubicación. También rocas o constituyentes de roca que han sido formados *in situ* (término creado por VON GÜMBEL en 1888) (opuesto: *alóctono*).

Autochtone: 1 – Unité tectonique formée à son emplacement actuel et n'ayant subi que des transports limités.
2 – Constituant de roche ou roche formée *in situ* (peu utilisé) (VON GÜMBEL, 1888) (Contraire: Allochtone).

Autochthon: Tektonische Einheiten oder Massen, die an der Fundstätte gebildet und nie merkbar transportiert worden sind. Auch: Gesteine oder Gesteinsbestandteile, die *in situ* entstanden sind (Ausdruck eingeführt durch VON GÜMBEL, 1888) (Gegenteil: Allochthon).

Azimuth

(of a line): The angle which a line forms with the true meridian as measured on an imaginary horizontal circle. The angle (which may be from zero to 360 degrees) is usually read clockwise from the north point. (MCKINSTRY)

Azimut (de una línea): El ángulo que forma una línea con el verdadero meridiano, medido en un círculo horizontal imaginario. El ángulo (que puede ser de cero a 360 grados) se lee generalmente en el sentido del reloj desde el punto norte.

Azimuth: L'azimuth d'une ligne est l'angle formé par un plan vertical passant par cette ligne et le méridien origine. Cet angle est lu à partir du Nord dans le sens des aiguilles d'une montre.

Azimuth (einer Linie): Der Winkel zwischen einer Linie und dem wahren Meridian, gemessen an einem imaginären Horizontalkreis. Der Winkel, der von 0 bis 360 Grad betragen kann, wird gewöhnlich im Uhrzeigersinn von der Nordrichtung aus gemessen.

Back:

The part of a vein or lode that is on the upper side of a drift or stope.

Techo: La parte de una veta o filón que está hacia la parte de arriba en un socavón o una labor minera.

Couronne, Extrados: Partie d'un filon ou d'un lode situé à la partie supérieure d'une galerie ou d'un gradin.

Firste: Das Erz oder Gestein, das eine Strecke oder einen Abbau, im weitesten Sinne jeden bergmännischen Hohlraum oben begrenzt.

Bailer:

A long cylindrical container with a valve at the bottom, which is used for removing the water, cuttings, and mud from the bottom of a cable-tool well. (LEVORSEN)

Achicador: Depósito cilíndrico largo con una válvula en el fondo, que es usado para sacar agua, menudencias o barro desde el fondo de un pozo de cable.

Cuiller, Ecope: Récipient cylindrique oblong, muni à son fond d'une soupape, utilisé pour épuiser l'eau, la boue ou les débris au fond d'un trou foré au trépan.

Sandfänger (Sandrohr): Ein länglicher zylindrischer Behälter mit einem Ventil am Boden, der zur Beseitigung von Wasser, Bohrklein und sonstigen Feststoffen vom Tiefsten einer Seilbohrung dient.

Banded vein:

Banded vein: See *vein*.

Veta bandeada: Ver veta.

Filon rubané: Voir filon.

Bändergang: siehe Gang.

Batholith (see Appendix VIIb):

An intrusive stock-like mass of igneous rock supposedly enlarging downward, and with a known diameter greater than 55 km (SCHMITT) or 110 km (LONGWELL, BATEMAN), (cf. *stock*).

Batolito: Gran masa de roca intrusiva que se supone ensanchada hacia abajo, con un diámetro conocido mayor de 55 kilómetros (SCHMITT) o 110 kilómetros (LONGWELL, BATEMAN), (cf. *stock*).

Batholite: Masse de roches intrusives en forme de stock dont on suppose qu'elle s'élargit en profondeur. Son diamètre connu doit être supérieur à 55 km (SCHMITT) ou à 110 km (LONGWELL et BATEMAN), (Voir *stock*).

Batholith: Eine intrusive Masse Gesteins, die sich angenommenerweise nach der Tiefe hin ausweitet und einen bekannten Durchmesser größer als 55 km (nach SCHMITT) oder 110 km (nach LONGWELL, BATEMAN) hat (s. auch *Stock*).

Bed:

(rock-stratigraphic & general): A bed is the smallest rock-stratigraphic unit recognized in classification (CSN). In popular usage any small unit, for example a coal seam of economic importance, may be named *bed*; this does not imply a formal stratigraphic connotation.

Capa o Estrato (lito-estratigráfica y general): Una capa es la mas pequeña unidad estratigráfica reconocida en clasificación. En lenguaje *general* cualquier unidad pequeña, por ejemplo, un estrato de carbón de importancia económica puede ser llamada *capa*; esto no supone una formal connotación estratigráfica.

Lit; Couche: Un lit est la plus petite unité lithostratigraphique reconnue dans la classification. En langage courant, toute petite unité, par exemple une couche de charbon d'importance économique peut être appelée lit; ceci ne comporte pas d'implication stratigraphique stricte.

Schicht (stratigraphisch und allgemein): Ist die kleinste stratigraphisch unterscheidbare Einheit. Allgemein jede kleine Einheit, z. B. ein Kohlenflöz von wirtschaftlicher Bedeutung; die Definition verlangt keine formelle stratigraphische Begriffsabgrenzung.

Bedded ore deposit:

See *ore, ore deposit*.

Depósito mineral estratificado: Ver mena, depósito mineral.

Gîte stratiforme: Voir minerai, gîte métallifère.

Schichtige Erzlagerstätte: siehe Erz, Erzlagerstätte.

Bedding, bedding plane:

Plane of stratification. The surface marking the boundary between a bed and the bed above or below it. (MCKINSTRY)

Estratificación, plano de estratificación: La superficie que determina el límite de una capa entre la de arriba y la de abajo.

Litage: Disposition finement stratifiée qui affecte l'intérieur des bancs et les séries qui ne sont pas divisées en lits et en bancs (LOMBARD). Stratification interne.

Plan de stratification: Surface constituant la limite entre un lit et le lit immédiatement supérieur ou inférieur.

Schichtung, Schichtebene: Ebene in der Ablagerung. Die Fläche, welche die Grenze einer Schicht gegenüber der darüber- oder darunter liegenden darstellt.

Biostratigraphic units:

A biostratigraphic unit is a body of rock strata characterized by its content of fossils contemporaneous with the deposition of the strata. (CSN)

Unidades bioestratigráficas: Una unidad bioestratigráfica es el cuerpo de un estrato rocoso caracterizado por su contenido de fósiles contemporáneos con la deposición del estrato.

Unité biostratigraphique: Ensemble de roches stratifiées caractérisé par les fossiles qu'il contient, dans la mesure où ceux-ci sont contemporains du dépôt des strates.

Biostratigraphische Einheit: Ein Komplex von Gesteinsschichten, der durch seinen (gleichaltrigen) Fossilinhalt gekennzeichnet ist.

Biostratigraphic zone:

A zone is the general basic unit in biostratigraphic classification. It is defined as a stratum or body of strata characterized by the occurrence of a fossil taxon or taxa from one or more of which it receives its name. (CSN.)

Zona bioestratigráfica: Una zona es la unidad básica general en clasificación bioestratigráfica. Es definida como un estrato o parte de un estrato caracterizado por un grupo o grupos de fósiles guía, de uno o más de los cuales recibe su nombre.

Zone biostratigraphique: La zone est l'unité fondamentale de la classification biostratigraphique. Elle est définie comme une couche ou un ensemble de couches caractérisé par la présence d'une espèce fossile ou de plusieurs espèces qui lui donnent son nom.

Biostratigraphische Zone: Eine Zone ist die allgemeine Grundeinheit in der biostratigraphischen Untergliederung. Sie ist definiert als eine Schicht oder eine Schichtenfolge, die durch das Vorkommen eines Fossils oder mehrerer Fossilien gekennzeichnet wird, nach dem oder nach denen auch die Bezeichnung erfolgt.

Bit:

The drilling tool that actually cuts the hole in the rock. Both cable-tool and rotary-tool bits are of various designs, depending on the kind of rock being drilled. Bit designs are patented. (LEVORSEN)

Broca: Herramienta de perforación que corta un hueco en la roca. Ambas brocas, de rotación y percusión, son de varios diseños de acuerdo a la roca en que va a emplearse. Muchas estan patentadas.

Outil: Outil de forage, qui, en fait, découpe le trou dans la roche. Les trépans (*cable tool bit*) comme les tricones (*rotary tool bit*) ont des caractéristiques diverses en fonction de la nature de la roche à forer. Beaucoup sont brevetés.

Bohrmeißel oder -krone: Bohrwerkzeuge, durch welche die eigentliche Abspanung oder Zertrümmerung des Gesteins bei der Herstellung eines Bohrloches erfolgt. Meißel für drehendes und schlagendes Bohren sind von verschiedenen Formen, je nach dem zu durchbohrenden Gestein. Viele Ausführungen sind patentiert.

Block caving:

See mining methods.

Hundimiento en bloque: Ver metodos de minería.

Blocs foudroyés: Voir méthodes minières.

Blockbau: siehe Abbaumethoden.

Bonanza:

An exceptionally rich shoot or bunch of ore, particularly with reference to gold and silver. (SCH.)

Bonanza: Una parte o un ramal de depósito excepcionalmente rico, particularmente con referencia al oro y la plata.

Bonanza: Terme espagnol désignant une colonne minéralisée ou un amas exceptionnellement riche. Il est en général réservé à l'or et à l'argent.

Erzfall: Eine außerordentlich starke Anreicherung von Erz, besonders bei Gold und Silber (Edelfall).

Branch
Offshoot; Apophysis: A branch of a vein or a dike to which it is attached. (Sch.)

Brazo: Ramal; apófisis — el brazo de una veta o dique al cual está unido.

Apophyse: Branche d'un filon ou d'un dike.

Ast, Abzweig, Apophyse, Trum: Verzweigung eines Ganges oder einer Ader.

Brecciated vein:
See vein.

Veta brechada: Ver veta.

Filon bréchoïde: Voir filon.

Brekziengang: siehe Gang.

Cable-tool drilling:
A method of drilling whereby a steel bit of varying design is fastened to a drill stem and jars, and the whole fastened on a wire line. When given an up-and-down motion, the bit cuts the rock by impact. The drill cuttings are removed by bailling them out after the tools have been removed from the hole. Formation waters and caving walls are controlled by setting a steel pipe, called *casing*, in the hole and sealing it in with cement. Drilling is then resumed with a smaller bit size. Holes range from seven to eighty cm in diameter, the larger sizes being near the surface and used for deep holes. (Levorsen)

Perforación a percusión: Un método de sondaje por el cual se coloca una corona de acero de variados diseños a una espiga y percutor, yendo el conjunto sujeto a un cable. Al darle un movimiento vertical la corona corta la roca por percusión. La roca molida se extrae mediante un achicador, después de haber levantado todas las herramientas de perforación. Filtraciones de agua y hundimiento de las paredes se controlan colocando tubos de acero, llamados *revestimiento* y sellando con cemento. Se continúa perforando con coronas más pequeñas. Los huecos fluctúan entre siete y ochenta cm de tamaño, estando los más grandes cerca de la superficie y siendo usados para huecos profundos.

Sondage au trépan (au cable): Méthode de sondage employant un outil de formes variées fixé à des tiges et masses-tiges, le tout étant attaché à un câble. Un mouvement alternatif vertical est imprimé à l'ensemble et l'outil entame la roche par percussion. Les débris sont remontés par écopage au moyen d'une cuiller, une fois tous les outils retirés du trou. Les eaux souterraines et la tenue des parois sont contrôlées par le *tubage* du trou et le scellement au ciment. Le sondage est ensuite repris avec un outil de dimension plus réduite. Les diamètres vont d'environ sept à quatre vingt cm les grandes dimensions étant proches de la surface et utilisées pour les trous destinés aux grandes profondeurs.

Seilschlagbohren: Ein Bohrverfahren, bei dem ein Stahlmeißel an einer Einheit aus Stahlgestänge, Schlagröhre und Bohrseil befestigt ist. Bei Auf- und Abbewegung zertrümmert der Meißel bei seinem Aufschlag auf die Bohrlochsohle das Gestein. Das Bohrklein wird mit Spezialbechern (Sandfänger) aus dem Bohrloch gehoben, nachdem die Schlageinheit entfernt worden ist. Eindringende Wässer und brüchige Bohrlochwandungen werden durch das Einbringen von Futterrohren beherrscht (sog. „casings"). Die Rohre werden in Zement eingegossen. Danach wird das Bohren mit jeweils kleinerem Meißeldurchmesser wieder aufgenommen. Die Bohrlochdurchmesser liegen zwischen sieben und achtzig cm, wobei die größeren Durchmesser nahe der Oberfläche eingesetzt und vorsorglich für tiefe Bohrungen gewählt werden.

Capping:

1. A rock formation (consolidated or unconsolidated) overlying a body of rock or ore; e. g., rhyolitic volcanics which form mesas and plateaus in S. W. U. S. and Mexico. 2. The oxidized equivalent of disseminated sulphide material. (LOCKE)

Cubierta: 1. Formación de roca (consolidada o inconsolidada) que descansa sobre un cuerpo de roca o depósito mineral; ejemplo: las volcánicas riolíticas, las cuales forman mesas y mesetas en el sur este de EE. UU. y México. 2. Equivalente oxidado de sulfuros diseminados.

Recouvrement: 1 – Formation (consolidée ou non) surmontant un ensemble rocheux ou un corps minéralisé. Ex.: Les roches volcaniques rhyolitiques qui forment des mesas et des plateaux dans les Etats-Unis du SW et le Mexique. 2 – Equivalent oxydé de sulfures disséminés (sens inusité en France).

—: 1. Eine Gesteinsformation (verfestigt oder unverfestigt), die einen Gesteins- oder Erzkörper überlagert; z. B. rhyolithische Erstarrungsgesteine bei den Tafelbergen und Plateaus im SW der USA und in Mexiko. 2. Das oxydierte Äquivalent zum Typ der Imprägnationslagerstätten (siehe Eiserner Hut).

Cap rock:

The covering or "capping" rock on top of an oil or gas reservoir or over a salt dome.

Roca de cubierta: Envoltura o cubierta de roca en el techo de un reservorio de petróleo o gas, o sobre un domo de sal.

—: Roche recouvrant le sommet d'un réservoir de pétrole ou de gaz ou encore d'un dôme de sel.

Deckgebirge: Das von der Tagesoberfläche bis zu einer Minerallagerstätte anstehende Gestein (Öl- oder Gaslagerstätte, Salzdom, Kohleflöz).

Casing:

Steel pipe of varying diameter and weight that comes in joints from five to eleven meters in length, which are joined together by threads and couplings at the well. Casing is "run" into the well hole for the purpose of supporting the walls of the well and preventing them from caving, for shutting off water either above or below the pay formation, and for shutting off portions of the pay formation such as shales, loose sands, or gas. The pay formations are opened up by perforating. (LEVORSEN)

Revestimiento: Tubos de acero de diámetro y peso variable que vienen en acoplamientos de cinco hasta once metros de largo, los cuales se unen mediante roscas y empalmes en el pozo. El revestimiento avanza dentro del sondaje con el objeto de sostener las paredes y prevenir su hundimiento, para desviar el agua arriba o abajo de la formación productiva y para interceptar porciones de la formación productiva, tales como lutitas, arenas movedizas o gas. Las formaciones productivas son exploradas mediante perforadoras.

Tubage de soutènement: Tuyaux d'acier de poids et de diamètre variables, livrés en tronçons de 5 à 11 m. de long, assemblés au moyen de filetages et de manchons avant la descente. Le tubage est glissé dans le puits afin de soutenir les parois et d'en empêcher l'éboulement, d'isoler les eaux au-dessus ou au-dessous de la formation productive et afin d'isoler certaines parties de la formation telles que des shales, des sables instables ou des gaz. Les formations productives sont libérées par perforation du tubage.

Futterrohre: Stahlrohre unterschiedlicher Durchmesser und Gewichte mit Einheitslängen zwischen 5 und 11 m, die beim Einhängen in ein Bohrloch durch Gewinde oder Kupplungen miteinander verbunden werden. Sie dienen zum Stützen der Bohrlochwandung, zum Abhalten von Formationswasser von oberhalb oder unterhalb des Förderhorizontes und zum Isolieren einzelner Partien von Förderhorizonten gegeneinander.

Catastrophism
(CUVIER, 1769—1832): The doctrine that changes in the earth's fauna and flora have been effected by catastrophic physical forces, followed by creation.

Catastrofismo (CUVIER, 1769—1832): Teoría según la cual los cambios en la fauna y flora de la tierra se deben a fuerzas físicas catastróficas, seguidos por creación.

Théorie des cataclysmes (CUVIER, 1769—1832): Théorie suivant laquelle les changements de faunes et de flores terrestres s'effectuèrent par des destructions catastrophiques et totales de la flore et de la faune, suivies par des créations.

Katastrophentheorie, Kataklysmentheorie (CUVIER, 1769—1832): Die Lehre, nach der durch Katastrophen Veränderungen in der Fauna und Flora hervorgerufen und neue Gattungen geschaffen worden seien.

Caving
See *mining methods*.

Hundimiento: Ver métodos de minería.

Foudroyage: Voir méthodes minières.

Bruchbau: Siehe Abbaumethoden.

Cement:
1. Material, which is placed in a drill-hole and forced in behind the casing in order to seal the casing into the walls of the hole and prevent any unwanted leakage of formation fluids into the well. 2. Chemically precipitated material, often mixed with fine detrital rock matter, occurring in the interstices between particles of clastic rocks and ores. Silica, carbonates, iron oxides and hydroxides, gypsum, and barite are the most common.

Cemento: 1. Material colocado y metido detrás del entubado de una perforación, a fin de sellar el revestimiento de las paredes del hueco y evitar goteras y filtraciones en el pozo. 2. Material formado por precipitación química, frecuentemente mezclado con productos detríticos finos, formando la matriz de las rocas y menas elásticas. (Sílice, carbonatos, óxidos e hidróxidos de hierro, yeso y baritina son los minerales más comunes).

Ciment: 1. Matériel placé dans le trou (de forage) et chassé derrière le tubage pour le sceller aux parois du sondage et empêcher toute fuite de fluide indésirable dans le puits. 2. Matériel formé par précipitation chimique, souvant mélangé avec de fins fragments détritiques, formant la matrice des roches et gîtes détritiques. (Silice, carbonates, oxydes et hydroxydes de fer, gypse et barytine sont les constituants les plus fréquents.)

Zement: 1. Material, welches ins Bohrloch hinter die Rohre gepumpt wird, um den Raum zwischen diesen und dem Gestein zu schließen, damit unerwünschtes Entweichen von Flüssigkeit unterbleibt. 2. Grundmasse: Chemisch ausgefälltes Material, oft gemischt mit feinem Gesteinsdetritus als Grundmasse in klastischen Gesteinen und Erzen. (Silikate, Karbonate, Eisenoxyde und -hydroxyde, Gips und Schwerspat sind die häufigsten Bestandteile.)

Cementation:
The process of transportation and deposition of mineral matter which precipitates as a binder between, or on the surface of rock grains. In common rocks, quartz, calcite, dolomite, siderite and iron oxide are frequent cementing materials. In the cementation zone of sulphide deposits (cf.) such sulphides as chalcocite are important cementing substances (see *cementation zone*).

Cementación: El proceso de transportación y deposición de la materia mineral, la cual precipita como aglutinante entre o sobre la superficie de los granos de la roca. En rocas comunes tales como cuarzo, calcita, dolomita, siderita y óxido de hierro, la cementación es frecuente. En la zona de cementación de depósitos de sulfuros, tales sulfuros, como la chalcocita son importantes sustancias de cementación (ver zona de *cementación*).

Cimentation: Processus de transport et de dépôt de la matière minérale qui précipite sous forme de liant entre les grains d'une roche ou à leur surface. Dans les roches courantes, le quartz, la calcite, la dolomite, la sidérose, et les oxydes de fer sont des ciments rencontrés fréquemment. On dit *cémentation* pour ce phénomène quand il s'agit de gîtes sulfurés. Dans les zones de cémentation des gîtes sulfurés, des sulfures tels que la chalcocite jouent également un rôle important de ciment (voir zone de *cémentation*).

Zementation: Der Vorgang des Transportes und der Ablagerung von Mineralsubstanzen in Form von Ausfällungen und Bindemittel zwischen oder an der Oberfläche von Gesteinspartikeln. Häufig sind Quarz, Calcit, Dolomit, Siderit und Eisenoxyde derartige Bindemittel. In der Zementationszone von Sulfiden ist Chalkosit eine wichtige Zementsubstanz (siehe *Zementationszone*).

Cementation zone (see zoning)
(of sulphide deposits): One of the supergene zones formed by weathering of an outcropping (or near surface) sulphide deposit; characterized by the cementative accumulation of material brought in, normally downward from the *oxidation* and *leaching* zones. This accumulation may or may not lead to a more valuable mineral deposit; common example of a more valuable and therefore "enriched" zone: the chalcocite zones of porphyry copper deposits. (See *cementation*.)

Zona de cementación (de depósitos de sulfuros): Una de las zonas supergenas formadas por alteración meteórica del afloramiento (o afloramiento cerca de la superficie) de un depósito de sulfuros; caracterizado por la acumulación cementativa de material introducido normalmente hacia abajo desde la zona de lixiviación y oxidación. Esta acumulación puede o no conducir a un depósito mineral de más valor; un ejemplo común de un depósito de más valor y por consiguiente de "zona de enriquecimiento" es la zona de chalcocita en los depósitos de cobre de tipo porfirítico. (Ver también cementación.)

Zone de cémentation: L'une des zones supergènes formées par l'altération d'un gîte sulfuré affleurant ou subaffleurant. Elle est caractérisée par l'accumulation de matériel par cémentation, à partir des zones d'oxydation et de lessivage. Cette accumulation peut conduire ou non à un gisement enrichi. Exemple courant d'enrichissement et de zone enrichie: les zones de chalcocite des porphyres cuprifères. (Voir cémentation.)

Zementationszone (sulfidische Lagerstätten): Eine supergene Zone, die durch Verwitterung an Ausbissen sulfidischer Lagerstätten entsteht; charakterisiert durch die zementative Ansammlung des normalerweise aus der oberhalb liegenden Oxydations- und Auslaugungszone beigeführten Materials. Diese Anhäufung kann, muß aber nicht, zu einer wertvolleren Minerallagerstätte führen; typisches Beispiel einer wertvollen und somit angereicherten Zone: Die Chalkosit-Zone von porphyritischen Kupferlagerstätten. (Siehe auch Zementation.)

Chambered vein:
See *vein*.

Veta cámara: Ver veta.

Filon en chambre: Voir filon.

Kammergang: siehe Gang.

Chorismatic
(or chorismites, polyschematic rocks and minerals): Fabric designation for a rock which consists clearly of units with different textures.

Corismático (corismitas o rocas y minerales poliesquemáticos): Denominación de fábrica de una roca, la cual consta claramente de unidades de diferentes texturas.

Texture chorismatique: Terme de structure désignant une roche composée de plusieurs unités ayant chacune une texture différente.

Chorismatisch (Chorismit, chorismatisches Gestein): Bauform eines Gesteins, das deutlich aus Einheiten verschiedener Strukturen besteht (z. B. Konglomerat, Brekzie).

Churn drilling
Cable-tool drilling, syn.

Perforación a percusión: Perforación a cable.

Sondage au cable: Voir sondage au trépan.

—: Seilschlagbohren.

Circle:
A term employed in the central U.S.A. for a more or less circular lead-zinc deposit developed in clayey chert breccias in old sinkholes in Paleozoic limestone or dolomite (broken ground). (SCH.)

13

Círculo: Término empleado en la parte central de U.S.A. para designar una forma más o menos circular de un depósito de plomo-zinc desarrollado en brechas de chert arcillosas en antiguos sumideros de calizas o dolomitas paleozoicas (terreno roto).

—: Terme employé dans le centre des Etats-Unis pour désigner un gisement de plomb-zinc plus ou moins circulaire placé dans des brèches argileuses à chert dans des puits d'effondrement traversant des calcaires ou dolomites paléozoïques (matériel fracturé).

—: Ein in der mittleren USA üblicher Ausdruck für eine mehr oder weniger kreisförmige Blei-Zink Lagerstätte, die in tonigen Hornstein-Brekzien in alten Hohlräumen (Dolinen) von paläozoischem Kalk oder Dolomit entwickelt ist.

Cockade ore:
See *ore*.

Mena en forma de cocarda: Ver mena.

Minerai en cocarde: Voir minerai.

Kokardenerz: siehe Erz.

"Collection recrystallization":
A type of re-crystallization in which a number of grains combine to form a larger grain (cf. *recrystallization*).

Cristalización por coalescencia: Tipo de recristalización en el cual un número de granos se redesarrollan para formar granos más grandes (ver recristalización).

Cristallisation par coalescence: Type de recristalisation dans laquelle un certain nombre de grains croissent à nouveau et aboutissent par coalescence à un cristal unique plus gros (voir recristallisation).

Sammelkristallisation: Eine Form der Rekristallisation, in der sich eine Anzahl Körner zu größeren Stücken weiterentwickelt (siehe Rekristallisation).

Colluvial placers (see Fig. 14):
Deposits formed by a combination of *diluvial* and *alluvial* processes (AGI) (cf. *placers*).

Placeres coluviales: Depósitos formados por la combinación de procesos diluviales y aluviales.

Placer colluvial: Gîte formé par une combinaison des processus alluviaux et diluviaux.

Colluviale Seifen: Lagerstätte, die durch Zusammenwirken von diluvialen und alluvialen Vorgängen entstanden sind.

Complex ores:
Ores from which several metals are recoverable.

Menas complejas: Menas de las cuales se recuperan diferentes metales.

Minerai complexe: Minerai donnant plusieurs métaux.

Komplexe Erze: Erze, aus denen mehrere Metalle gewonnen werden können.

Composite section:

Projection of data from various locations into a single vertical (or inclined) section. (McKinstry)

Sección compuesta: Proyección de datos de varios sitios en una simple sección vertical (o inclinada).

Section renseignée: Projection de renseignements provenant de divers points sur une section unique verticale (ou inclinée).

Sammelriß: Projektion von Daten verschiedener Punkte in ein einzelnes vertikales oder geneigtes Profil.

Concentrate:

Verb: To separate valuable minerals from the associated gangue or rock. Noun: Final product of the separation of (most) of the valuable ore minerals from (most) of the gangue material (opp. *tailings*).

Concentrar: Verbo: Separar los minerales valiosos de la ganga o roca asociada. Sustantivo: Concentrado: El producto final de la separación de (la mayor parte de) los minerales valiosos de una mena (de la mayor parte) del material de la ganga (opuesto de colas).

Concentré (verbe: concentrer, valoriser): Résultat de la valorisation (ou du traitement) du minerai tout venant, dans lequel tout (ou la majeure partie) du métal ou des métaux recherchés a été concentré et toute (ou une partie de) la gangue a été éliminée (contraire: stérile).

Konzentrat: Endprodukt der Trennung der wertvollen Minerale eines Erzes vom begleitenden Ganggestein (Gegenteil: Abgänge oder Berge).

Concentrator:

Plant ("*mill*") in which the ore minerals are concentrated.

Concentradora: Planta ("molino") en la cual los minerales valiosos son concentrados.

Usine de traitement de minerai: Usine où le minerai est valorisé. Quelquefois on la désigne par le terme impropre de laverie ou de lavoir.

Aufbereitung: Anlage oder Vorgang, wodurch Erzmineralien konzentriert und Abgänge (Berge) ausgeschieden werden. (siehe Konzentrat.)

Congruent (congruence, congruency) (see Fig. 1):

Superposable, so as to be geometrically coincident throughout.

Congruente (congruencia): superponible; relación entre dos formas superpuestas, coincidentes geométricamente.

Congruent (congruence): Relation entre deux formes superposables de manière à coïncider géométriquement dans toute leur étendue.

Kongruent (Kongruenz): – –; Bezeichnung für geometrische Übereinstimmung durch Überlagerung.

Connate water:

The original pore water trapped during diagenetic crystallization and kept in place ever since.

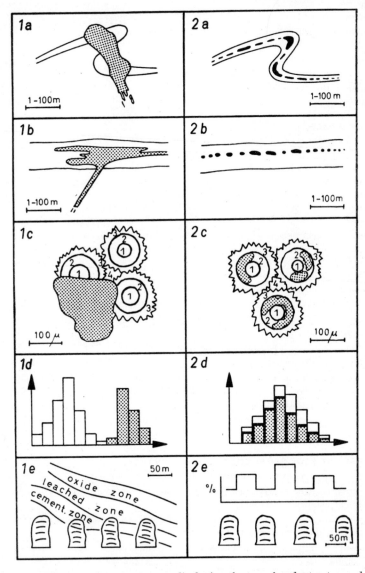

Fig. 1. Five different geometric patterns displaying the two abundant extreme degrees of congruency (2a, 2b, . . .) and non-congruency (1a, 1b, . . .).

Agua congénita: El original poro de agua atrapado durante la cristalización diagenética y mantenido desde entonces en su lugar.

Eau connée: Eau interstitielle enfermée durant la cristallisation diagénétique et conservée dans la roche depuis lors.

—: Das ursprünglich bei der Diagenese in den Poren eingeschlossene und seither darin verbliebene Wasser.

Contact metamorphism (see Fig. 6a):

Changes of country rock or of mineral deposits that take place near an intrusion, with or without addition or subtraction of mineral matter.

Metamorfismo de contacto: Cambios con o sin adición o/y sustracción, con calor y acción química en una roca de caja invadida por una intrusión.

Métamorphisme de contact: Changements dans la composition physique et chimique d'une roche encaissante ou d'un gîte au voisinage d'une intrusion, pouvant avoir lieu avec départ ou apport de matière.

Kontaktmetamorphose: Umwandlung eines Gesteins, in der Nähe einer Intrusion mit oder ohne gleichzeitigem Materialaustausch.

Contact metasomatism (see Fig. 6a):

A change through addition and/or subtraction, heat and chemical action, in a country rock invaded by an intrusion.

Metasomatismo de contacto: Cambio por adición y/o sustracción más calor y acción química en una roca de caja invadida por una intrusión.

Métasomatose de contact: Modification par addition ou (et) soustraction, par la chaleur et une action chimique, d'une roche envahie par une intrusion.

Kontaktmetasomatose: Eine Änderung durch Zu- oder Abfuhr, durch Temperatur und chemische Reaktionen in einem Gestein, das von einer Intrusion durchdrungen wird.

Contact vein:

See *vein*.

Veta de contacto: Ver veta.

Filon de contact: Voir filon.

Kontaktgang: siehe Gang.

Countervein:

See *vein*.

Veta contraria: Ver veta.

Filon croiseur: Voir filon.

Quergang: siehe Gang.

Country rock:

The common rock of an area or a local place, containing ore matter.

Roca de caja: La roca común de un área o un lugar dado que contiene materias minerales económicas, singenéticas o epigenéticas.

Roche encaissante: Roche commune dans une région ou un endroit donné et contenant des minéralisations syngénétiques ou épigénétiques.

Nebengestein: Das ein Mineralvorkommen unmittelbar umgebende Gestein, in das Erzmineralien syngenetisch oder epigenetisch eingebettet sind.

Crosscut:

A level driven across the course of a vein, or in general across the direction of the main workings, or across the "grain of the coal". (RAYMOND, via FAY)

Crucero, cruzados: Socavón transversal al curso de una veta o en general transversal a la dirección de los trabajos principales o tranversal a la "veta de carbón".

Travers banc: Galerie creusée à travers un filon ou en général à travers la direction des travaux principaux ou à travers le «grain du charbon».

Querschlag: Ein bergmännischer Hohlraum (Strecke), der rechtwinklig (querend) zum Streichen einer Lagerstätte oder rechtwinklig zur Hauptstrecke aufgefahren ist. Die Strecke durchquert verschiedene Schichten oder Flöze (Gegenteil: Richtstrecke).

Crustified vein:

See *vein*.

Veta costrificada: Ver veta.

Filon crustifié: Voir filon.

Krustenartiger oder kokardenartiger Gang: siehe Bändergang, Brekziengang, Kokardenerz.

Crystallization differentiation (see Fig. 5 and Appendix VIIIa, b, c):

See *magmatic differentiation*.

Diferenciación por cristalización: Ver diferenciación magmática.

Différenciation par cristallisation: Voir différenciation magmatique.

Kristallisationsdifferentiation: Siehe magmatische Differentiation.

Descension theory

(WERNER and neptunists, 18th century): The theory that the material in veins was introduced in solution from above. (SCH.)

Teoría descensional (WERNER y neptunistas, siglo XVIII): Teoría según la cual el material de las vetas ha sido introducido en solución desde arriba.

Descensionnisme (WERNER et les Neptunistes, XVIIIème siècle): Théorie selon laquelle le matériel remplissant une caisse filonienne a été introduit par des solutions venant du haut.

Deszendenztheorie (WERNER und Neptunisten, 18. Jh.): Theorie, nach der das Gangmaterial durch Lösungen von oben her eingeführt worden ist.

Deuteric

(= autohydrothermal): Material or processes which are hydrothermal in nature, but restricted to the source rock itself (e. g., deuteric alteration, etc.) and to the late stages of magmatic crystallization, not including any later changes.

Deutérico (= autohidrotermal): Material o procesos de naturaleza hidrotermal, pero restringidos al origen de la roca misma (ejemplo, alteración deutérica, etc.) y al último estado de la cristalización magmática, sin incluir cualquier cambio posterior.

Deutérique (= autohydrothermal): Matériel ou processus de nature hydrothermale, mais limité à la roche mère elle-même (ex. altération deutérique) et aux phases tardives de la cristallisation magmatique et n'incluant pas les modifications ultérieures.

Deuterisch (= autohydrothermal): Bezeichnungen für Umwandlungen, die durch hydrothermale Restlösungen der magmatischen Erstarrung am bereits kristallisierten Material selbst stattgefunden haben, in direkter Fortsetzung der Erstarrung selbst.

Develop:

The act of working ore bodies in mines, as by shaft sinking, drift driving and rock removal (see also next term).

Desarrollar: El acto de trabajar un depósito mineral en las minas, como un pique, un socavón y una remoción de roca (véase el próximo término).

Préparer: Préparer l'exploitation d'amas minéralisés par une mine en fonçant des puits, en creusant des galeries et en enlevant des stériles (voir le terme suivant).

Ausrichten: Herstellen von Grubenräumen in Verbindung mit der Ausrichtung (vergleiche den nächsten Ausdruck).

Development work:

Work undertaken to open up orebodies as distinguished from the work of actual ore extraction. Sometimes development work is distinguished from exploratory work on the one hand and from stope preparation on the other. (MCKINSTRY)

Trabajo de desarrollo: Trabajo emprendido a fin de hacer accesible el cuerpo mineral, distinto del trabajo real de extracción del mineral. A veces el trabajo de desarrollo se distingue, por una parte, del trabajo exploratorio, y por otra, del trabajo de preparación de galerías de extracción.

Travaux préparatoires: Travaux entrepris pour ouvrir des corps minéralisés se distinguant de l'exploitation véritable. Quelquefois on distingue les «travaux préparatoires» des travaux de reconnaissance d'une part et de la préparation des chantiers d'autre part.

Ausrichtung: Die Herstellung von Grubenräumen, die dazu dient, eine Lagerstätte für den Abbau zu erschließen (z. B. Schachtabteufung, Streckenvortrieb u. a.). Die Ausrichtung läuft der Vorrichtung (vorbereitende Arbeiten zur unmittelbaren Inangriffnahme eines Abbaues) und der Gewinnung (eigentlicher Abbau, Lösen aus dem natürlichen Verband) voraus.

Diagenesis:

Diagenesis designates the process of consolidation of a sediment, embracing the time from its deposition to the time at which it has become a hard sedimentary rock. Examples are: consolidation (lithification, induration) of calcareous ooze to limestone, clay-ooze to mudstone, clay, or shales, and of sands to sandstones. (Term introduced by VON GÜMBEL, 1868).

Diagenesis: Se denomina diagenesis el proceso de consolidación de un sedimento, comprendiendo el tiempo desde su deposición hasta el tiempo en el cual se convierte en una roca sedimentaria dura. Ejemplo: consolidación (litificación, induración) de cieno calcáreo en caliza, cieno arcilloso en lodolita, arcilla o lutita, y de arenas en arenisca. (Término introducido por VON GÜMBEL en 1868.)

Diagénèse: Processus de consolidation des sédiments, s'étendant depuis leur dépôt jusqu'au moment où ils sont transformés en une roche dure. Ex.: consolidation (lithification, induration) de boues calcaires en calcaire, de boues argileuses en argillites, argiles ou shales et de sables en grès. (Terme introduit par VON GÜMBEL en 1868).

Diagenese: Der Verfestigungsvorgang bei einem Sediment, vom Zeitpunkt der Ablagerung bis zu dem, wo es ein hartes Sedimentgestein geworden ist. Beispiele: Verfestigung (Konsolidation) von kalkigem Schlamm zu Kalkstein, von Letten zu Schieferton, Ton oder Mergel, oder von Sand zu Sandstein. (Ausdruck durch VON GÜMBEL 1868 eingeführt.)

Diagenetic differentiation:

Separation and migration of solid and fluid phases during diagenetic crystallization.

Diferenciación diagenética: La separación y migración de fases y soluciones durante la cristalización diagenética.

Différenciation diagénétique: Séparation et migration de phases solides et liquides durant la cristallisation diagénétique.

Diagenetische Differentiation: Absonderung und Wanderung von festen und fluiden Phasen während der diagenetischen Kristallisation.

Diaphthoresis:

Retrogressive or retrograde metamorphism. The process by which metamorphic minerals of a lower grade are formed at the expense of minerals which are characteristic of a higher grade of metamorphism. (SCH.)

Diaftoresis: Regresión o metamorfismo retrógrado. El proceso por el cual los minerales metamórficos de bajo grado se forman a expensas de minerales que son característicos de alto grado de metamorfismo.

Diaphtorèse: Métamorphisme rétrograde. Processus par lequel des minéraux métamorphiques d'un niveau inférieur sont formés aux dépens de minéraux caractéristiques d'un niveau de métamorphisme plus intense.

Diaphtorese: Retrogressive oder retrograde Metamorphose. Ein Prozess, bei dem metamorphe Minerale niederen Grades auf Kosten von Mineralen gebildet werden, die ihrem Charakter nach einer höheren Stufe der Metamorphose entsprechen.

Dike

(dyke): In general a crosscutting tabular fissure filling in rocks (consisting of sedimentary or igneous rocks and ores, from ascending or descending solutions, or lateral secretion). (opposite: sill).

Dique: En general una masa tabular llenando fisuras (consiste de roca o mena sedimentaria o ignea, descendente, ascendente o de secrecion lateral). (op.: sill).

Dike; filon: Masse tabulaire de roches remplissant une fissure (consiste en roches ou minerai sédimentaires ou ignées, per ascensum, ou descensum, ou par sécrétion latérale). (contraire: sill).

Gang: Meist plattenförmige Ausfüllungen von Spalten im Gestein (durch magmatische und sedimentäre Gesteine oder Erze, aszendent, deszendent oder lateralsekretionär).

Diluvial placers:

Material accumulated by flood action (rather than only by normal streams).

Placeres diluviales: Material acumulado por acción de inundaciones (más que por corrientes normales).

Placer diluvial: Matériel accumulé par des crues (plutôt que par des fleuves normaux seuls).

Diluviale Seifen: Durch Überflutungen (eher als durch regelmäßige Strömungen) angesammeltes Material.

Dip:

The inclination of a bed, vein, fault, etc., measured from the horizontal, thus the angle between a line in the bed perpendicular to the strike and the horizontal plane. (McKinstry)

Buzamiento: La inclinación de una capa, veta, falla, etc., medida desde la horizontal; es, pues, el ángulo entre una línea en la capa perpendicular al rumbo y un plano horizontal.

Pendage: Angle entre le plan horizontal et la ligne de plus grande pente d'un plan (strate, filon, faille, etc.).

Fallen, Einfallen: Neigung von Schichten, Gängen, Störungsflächen, Klüften, etc., gemessen von der Horizontalebene zu jener Linie der betrachteten Ebene, die rechtwinklig zum Streichen derselben liegt.

Directional drilling:

The controlled drilling of deflected holes by special orientation surveys. (Levorsen).

Perforación directriz: La perforación de desviación controlada mediante dispositivos especiales de orientación.

Sondage dirigé: Forage contrôlé de trous déviés par des méthodes topographiques spéciales.

Gerichtetes Bohren: Das Bohren mit geplanter und kontrollierter Ablenkung der Bohrlochrichtung mit Hilfe spezieller Richt- und Meßgeräte.

Drag:

1. Fragments of ore torn from a *lode* by fault movement and remaining in the fault zone. 2. Distortion of beds adjacent to a fault. (McKinstry).

Arrastre: 1. Fragmentos de mineral arrancados de una veta por un movimiento de falla y que permanecen en la zona de falla. 2. Distorsión de capas adyacentes a una falla.

—: 1 — Fragments de minerai arrachés à un lode par un mouvement de faille et restant dans la zone de faille. 2 — Distorsion de lits adjacents à une faille (= Rebroussement).

—: 1. Bruchstücke von Erz, die durch tektonische Bewegung aus einem Gang gebrochen wurden und in der Verwerfungszone verblieben sind. 2. Verzerrung von Schichten, die an eine Verwerfung stoßen.

Drag ore:

See *ore*.

Mena arrastrada: Ver mena.

Minerai entraîné: Voir minerai.

Abgeschertes Erz: siehe Erz.

Drift (see Appendix III):

A horizontal underground passage following or parallel to a vein. It is distinguished from a crosscut, which intersects a vein, or a level or gallery, which may either follow or intersect the vein. (Raymond via Fay)

Socavón: Pasaje horizontal subterráneo que sigue o corre paralel a una veta. Se diferencia de un crucero, en que éste intersecta una veta, o de un nivel o una galería en que éstos pueden o no seguir o intersectar una veta.

Traçage: Passage horizontal souterrain suivant ou parallel à un filon. On le distingue du traversbanc qui recoupe un filon ou d'une voie ou niveau qui peut, soit suivre, soit recouper un filon.

Strecke: Ein horizontaler, langgestreckter Grubenbau, parallel oder entlang dem Streichen eines Erzganges. Siehe auch Querschlag, Richtstrecke, Gangstrecke, Flözstrecke, Füllort, Umtrieb und Durchhieb.

Drift

(glacial): All the moraine material in transport by glacier ice, and all the material predominantly of glacial origin deposited directly by glaciers, or indirectly in glacial streams, glacial lakes, or the sea.

Escombro (glaciar): Todo el material transportado por un glaciar (morenas) y todo el material predominantemente de origen glacial, depositado directamente por glaciares o indirectamente en corrientes glaciales, lagos glaciales y el mar.

Moraine: Tout le matériel transporté par un glacier et tout le matériel d'origine surtout glaciaire, déposé directement par les glaciers ou indirectement dans des fleuves et lacs glaciaires et la mer.

Geschiebe, Gletschergeschiebe: Durch Gletscher bewegtes Moränenmaterial und solches, das vorwiegend — direkt oder indirekt — von Gletscherbewegung stammt oder in Gletscherströmen, -seen und im Meer abgelagert wurde.

Drilling mud:

The watery mud pumped down the drill pipe, out through the bit, and back up in the annular space between the drill pipe and the walls of the well, in rotary drilling. The mud carries the cuttings up from the bottom and sheathes the walls of the holes with a mud cake that prevents the walls from caving and keeps out formation water. The weight of the mud is regulated by adding various substances so that it is greater than the formation pressures expected. (LEVORSEN).

Barro de perforación: El barro acuoso se bombea a través de los tubos hasta la corona y se devuelve en el espacio anular entre los tubos y las paredes del hueco, en la perforación rotaria. El barro trae los detritos de perforación desde el fondo y reviste las paredes del hueco con una capa que previene los desprendimientos y permeabiliza las formaciones de agua. El peso del barro es regulado agregando varias substancias para que sea mayor que las presiones que se esperan.

Boue de sondage: Boue liquide pompée dans les tiges de sondage, traversant l'outil et remontant dans l'espace annulaire entre tiges et tubage dans le sondage rotary. La boue remonte les débris du fond et revêt les parois du trou d'une couche de boue et de ce fait les soutient et empêche les eaux de les traverser. Le poids de la boue est modifié par l'adjonction de diverses substances de façon à ce qu'il soit supérieur aux pressions prévisibles des terrains.

Bohrspülung: Spülflüssigkeit, die durch das Bohrgestänge nach unten und im Ringraum zwischen der Bohrlochwandung und dem Bohrgestänge hoch- oder umgekehrt gepumpt wird. Sie dient dem Abtransport des Bohrkleins, zur Kühlung und Reibungsverminderung, zum Aufbau eines Gegendruckes gegen Formationswasser und Gase, sowie zur Stützung der Bohrlochwandung. Ihre spezifischen Eigenschaften werden durch besondere Zuschläge so beeinflußt, daß der Druck von Gas und Wasser in den einzelnen Horizonten beherrscht werden kann.

Drill pipe:

A string of steal pipes, screwed together and extending from the rig floor to the drill collar and bit at the bottom of the hole. The drill pipe transmits the rotating motion from the derrick to the bit and conducts the drilling mud (q. v.) from the surface to the bottom of the hole. (LEVORSEN).

Tubería de perforación: Conjunto de tubos atornillados que se extienden desde la superficie a la boca de la perforación y hasta el fondo de la misma. La tubería transmite el movimiento de rotación de la máquina a la corona y conduce el barro de perforación de la superficie al fondo de la perforación.

Tige de sondage: Ligne de tubes d'acier assemblés par filetage et manchons, s'étendant de la plateforme de travail du chevalement au niveau du sol et se prolongeant jusqu'à l'outil au fond du sondage. Elle transmet le mouvement à l'outil et conduit les boues de sondages.

Bohrgestänge: Ein Strang von Stahlstangen oder -rohren, die – zusammengeschraubt – den Meißel bzw. die Krone mit dem Antrieb übertage verbinden. Beim drehenden Bohren überträgt das Gestänge die Drehbewegung vom Antrieb zur Bohrkrone und leitet darüber hinaus die Spülung zum Bohrlochtiefsten.

3*

Druse:

A small cavity lined with "drusy" minerals, usually the same as those that constitute the enclosing rock. Also: Central opening left between the last layers of minerals deposited in a cavity filling. Generally the same as *geode*.

Drusa: Una pequeña cavidad, revestida con racimos de minerales, generalmente los mismos que aquellos que constituyen la roca encajonante. También la abertura central dejada entre las últimas capas de minerales depositados en una cavidad de relleno. Generalmente es sinónimo de *geoda*.

Druse: Petite cavité tapissée de minéraux «de druse», généralement les mêmes que ceux qui constituent les roches encaissantes. Egalement: ouverture centrale laissée entre les dernières couches de minéraux déposés dans un remplissage de cavité. Généralement le même sens que *géode*.

Druse: Ein kleiner Hohlraum belegt oder ausgefüllt mit „Drusenmaterial", meist dieselben Minerale wie die des Nebengesteins. Auch häufig die Öffnung selbst, die zwischen den letzten Lagen von angelagertem Material in einem solchen Hohlraum frei geblieben ist. Im allgemeinen gleichbedeutend mit *Geode*.

Echelon

(*en échelon*): An arrangement of faults, veins, etc., in which the individuals are staggered like the treads of a staircase. (McKinstry)

Escalón (*en échelon*): *Disposición* de fallas, vetas, etc., en el cual los individuos están alternados como los peldaños de una escalera.

Echelon, *en échelon*: Disposition de failles ou de filons, telle que sur une coupe on voit une figure ressemblant aux barreaux d'une échelle.

Echelon (*en échelon*): Eine Anordnung von Verwerfungen, Gängen etc., in denen die einzelnen Stufen auseinandergezerrt sind wie Treppen einer Hausstiege.

Edelfall:

A German term for a shoot of precious metal ore. (Sch.)

—: Término alemán para designar un ramal de una mena de metal precioso.

—: Terme allemand désignant une colonne minéralisée en métaux précieux.

Edelfall: Anhäufung von Edelmetallerzen (siehe auch Erzfall).

Edle Geschicke:

A German term for rich silver ores from the intersection of vein systems in the Freiberg area. (Sch.)

—: Término alemán para designar ricas menas argentíferas en la intersección de vetas, en el área de Freiberg.

—: Terme allemand désignant les minerais d'argent riches des intersections filoniennes de Freiberg.

Edle Geschicke: Reiche Silbererze, die in den Schnittlinien von Gängen im Freiberger Revier vorkommen.

Eluvial placers (see Fig. 14):

Placer minerals moved and sometimes concentrated on talus slopes descending from the original deposit.

Placeres eluviales: Minerales de placeres transportados y a veces concentrados en los taludes de escombros descendentes desde el depósito original.

Placer éluvial: Minéraux de placers déplacés et quelquefois concentrés sur des pentes de talus s'abaissant en-dessous du gisement primaire.

Eluviale Seifen: Abgelagertes Material, das erneut transportiert und manchmal unterhalb der ursprünglichen Lagerstätte in Schutthalden angehäuft wurde.

Endogenous, endogenetic, endogenic:

Produced from within originating from, or due to internal causes. (WEBSTER) (To be applied in a basic way to any material or process, and therefore not to be used as a synonym of infra- or intracrustal, cf.); (opp. *exogenous, exogenetic, exogenic*).

Endógeno, endogenético, endogénico: Originado en el interior o debido a causas internas. (Debe ser aplicado en un sentido básico a cualquier material o proceso y por lo tanto no debe usarse como sinónimo de infra o intracortical, (ver estas voces); (opuesto: exógeno, exogenético, exogénico).

Endogène: Produit de l'intérieur, ou dû à des causes internes. (Doit être utilisé fondamentalement pour désigner des matériaux ou des processus et ne doit donc pas être utilisé comme synonyme d'infra- ou intracrustal, voir ces mots); (contraire: exogène).

Endogen, Endogenetik, endogenetisch: Von innen herkommend. Entstanden durch Vorgänge im Inneren. (Kann grundsätzlich für jedes Material oder jeden zutreffenden Vorgang verwendet werden und soll daher nicht als Synonym zu infra- oder intrakrustal gebraucht werden, siehe dort); (Gegenteil: exogen, Exogenetik, exogenetisch).

En echelon

(En échelon): See *echelon*.

En échelon: Ver escalón.

En échelon: Voir échelon.

En échelon: siehe Echelon.

Eolian placers:

Mineral grains accumulated by wind.

Placeres eólicos: Granos de mineral acumulados por el viento.

Placer éolien: Grains minéraux accumulés par le vent.

Aeolische Seifen: Mineralien, die durch Windbewegung angereichert wurden.

Epigenetic

(with regard to rocks and mineral deposits): Epigenetic are those rocks, mineral deposits, textures, structures, processes, etc. which formed later than the enclosing rock(s), as contrasted to *syngenetic* rocks, minerals, etc. which formed contemporaneously with the enclosing rocks.

Epigenético (con relación a rocas y depósitos minerales): Epigenéticos son aquellas rocas, depósitos minerales, texturas, estructuras, procesos, etc., que han sido formados posteriormente a la roca o rocas de caja, en contraste con rocas, minerales, etc. *singenéticos*, los cuales se han formado contemporáneamente a la roca de caja.

Epigénétique (en parlant de gîtes minéraux): Roches, gîtes, minéraux, textures, structures, processus, etc. développés après la formation des roches encaissantes et s'opposant ainsi aux roches et minéraux *syngénétiques*, qui se formèrent en même temps que les roches encaissantes.

Epigenetisch (bezogen auf Gesteins- und Minerallagerstätten): Epigenetisch sind Gesteine, Minerallagerstätten, Strukturen, Texturen, Vorgänge etc., die später entstanden sind als das einschließende Gestein; im Gegensatz zu *syngenetischen* Gesteinen, Mineralien etc., die gleichzeitig mit dem Nebengestein gebildet wurden.

Epithermal (see Appendix VIIb):

Hydrothermal processes, or minerals presumably formed at relatively low temperatures (cf. *meso-* and *katathermal*).

Epitermal: Procesos hidrotermales o minerales presumiblemente formados a temperatura relativamente baja (ver mesotermal y catatermal).

Epithermal: Minéraux ou processus hydrothermaux se développant à une température supposée basse (voir méso- et catathermal).

Epithermal: Hydrothermale Prozesse und Mineralbildungen, die bei relativ niedrigen Temperaturen stattfinden (siehe meso- und katathermal).

Exhalation:

1. An emanation or sending forth, as in the form of steam or vapor; that which is exhaled. (WEBSTER). 2. In geology, an emanation of gases or vapors, ordinarily formed beneath the surface of the earth, escaping from a volcanic conduit or fissure, or from molten lava, or from a fumarol or hot spring. (WSI.)

Exhalación: 1. Emanación o emisión, en forma de vaho o vapor (WEBSTER). 2. En geología, emanación de gases o vapores, comunmente formada debajo de la superficie de la tierra, que se escapa desde un conducto o fisura volcánica o desde una lava fundida, fumarola o fuente de agua termal.

Exhalaison: 1 – Emanation ou émission sous forme de vapeur; ce qui est exhalé. 2 – En géologie: émanations de gaz ou de vapeurs, formés habituellement en-dessous de la surface de la terre, et s'échappant d'une fissure ou d'un évent volcanique, de laves fondues, de fumerolles ou d'une source chaude.

Exhalation: 1. Emanation in Dampf- oder Gasform; dasjenige, was exhaliert wird. 2. In der Geologie Ausbrüche von Gasen und Dämpfen, die gewöhnlich unter der Erdoberfläche angereichert werden und durch vulkanische Kanäle, Bruchstellen, an der Oberfläche flüssiger Laven, durch Fumarolen oder heiße Quellen entweichen.

Exhalative deposits (see Appendix VIIb):
See volcanic exhalative mineral deposits.

Depósitos exhalativos: Ver depósitos minerales exhalativos volcánicos.

Gîtes exhalatifs: Voir gîtes volcaniques exhalatifs.

Exhalative Lagerstätten: Siehe vulkanisch exhalative Lagerstätten.

Exogenous, exogenetic, exogenic:
Produced from without; originating from, or due to external causes. (WEBSTER) (To be applied in a basic way to any material or process, and therefore not to be used as a synonym of supracrustal, cf.); (opp. *endogenous, endogenetic, endogenic*).

Exógeno, exogenético, exogénico: Originado desde afuera; origen procedente o debido a causas externas. (Debe ser aplicado en un sentido básico a cualquier material o proceso y por lo tanto no debe usarse como sinónimo de supracortical, ver esta voz); (opuesto: endógeno, endogenético, endogénico).

Exogène: Produit de l'extérieur, ou dû à des causes externes. (Doit être utilisé fondamentalement pour désigner des matériaux ou des processus et ne doit donc pas être utilisé comme synonyme de supracrustal, voir ce mot); (contraire: endogène).

Exogen, Exogenetik, exogenetisch: Entstanden von außen her; von äußeren Ursachen stammend. (Grundsätzlich für jeden Prozeß, jedes Material oder jeden zutreffenden Vorgang zu verwenden, kann aber nicht als Synonym für suprakrustal gelten, siehe dort); (Gegenteil: endogen, Endogenetik, endogenetisch).

Exsolution minerals:
Minerals that form through exsolution from other minerals by cooling and are mostly included in them. (SCH.)

Minerales de exsolución: Minerales que se forman por exsolución a partir de otros minerales por enfriamiento y generalmente en inclusión en estos últimos.

Minéraux d'exsolution: Minéraux formés par exsolution à partir d'autres minéraux lors du refroidissement et généralement en inclusion dans ces derniers.

Entmischungsminerale: Minerale, die durch Entmischung infolge Abkühlung aus anderen Mineralien geformt werden und meist in diesen eingeschlossen bleiben.

Fabric:
The whole of all geometric features or properties of an aggregate; includes thus texture and structure.

Fábrica: Conjunto de todos los caracteres o propiedades geométricas de un agregado; engloba, pues, textura y estructura.

Structure: Ensemble des caractères ou propriétés géométriques d'un agrégat; englobe donc texture et structure. N.B. – L'emploi du mot fabrique s'institue en français.

Gefüge: Gesamtheit der geometrischen Merkmale oder Eigenheiten eines Aggregats; umfaßt Struktur und Textur.

Face:

In any adit, tunnel, or stope, the end at which work is in progress or was last done. (FAY)

Frente: En cualquier socavón, túnel o galería, la parte final de un trabajo que está en curso de ejecución o que fue hecho al final.

Front: Dans une galerie, un tunnel ou un gradin, extrémité où s'est effectué en dernier le travail, ou est encore en cours d'exécution.

Ort, Stoß: Die Stelle in einem Grubenraum, an der Erweiterungsarbeiten verrichtet werden, wurden oder zu verrichten sind.

Facing

(of strata): Direction in which the original top of a vertical or steeply dipping bed now faces. (MCKINSTRY)

Cara (de un estrato): Dirección en la que mira actualmente la parte superior originaria de una capa vertical o muy inclinada.

Regard (d'une couche): Direction dans laquelle regarde actuellement le sommet original d'un lit vertical ou à fort pendage.

—: Richtung, in der die ursprünglich obere Fläche einer vertikalen oder steil stehenden Schicht zum Zeitpunkt der Untersuchung liegt.

Fahlband:

A Scandinavian mining term for metamorphic rocks heavily impregnated with finely divided sulphide minerals which in weathered condition give a brownish color to the country rock. (SCH.)

Fahlband: Término minero escandinavo para rocas metamórficas impregnadas fuertemente con minerales de sulfuros, los cuales, bajo la acción de la intemperie, dan un color castaño a la roca encajonante.

Fahlband: Terme minier scandinave désignant des roches métamorphiques fortement imprégnées par des sulfures finement divisés qui donnent lors de l'altération une couleur brune à la roche encaissante.

Fahlband: Ein skandinavischer Bergbauausdruck für metamorphe Gesteine, die stark mit fein verteilten Sulfidmineralien imprägniert sind; geben im verwitterten Zustand dem Nebengestein meist eine bräunliche Färbung.

Fault (see Appendix IV, V):

A fracture along which there has been displacement (ranging anywhere from a few cm to several km), (compare *fracture*, *joint* and *shear*).

Falla: Fractura en el plano de la cual ha habido desplazamiento (desde unos cms. hasta kms.), (ver fractura, juntura o diaclasa y cizallamiento).

Faille, rejet: Une fracture le long de laquelle il y a eu un déplacement (de l'ordre de quelques cm à quelques km), (voir fracture, diaclase et cisaillement).

Verwerfung, Sprung: Ein Bruch, längs welchem Verschiebungen stattgefunden haben (von wenigen cm bis km), (vergleiche Bruch, Abtrennungsflächen und Scherklüfte).

Fill, waste:

Valueless material rejected during mining or concentration, and placed in mined out stopes of a mine.

Relleno: Material sin valor usado para llenar labores explotadas de una mina.

Remblai: Matériau sans valeur servant a remplir les parties exploitées d'une mine.

Versatz, Berge, z. T. Alter Mann: Wertloses Gesteinsmaterial mit welchem bereits abgebaute Teile einer Grube gefüllt werden.

(als *Alter Mann* werden auch die planmäßig zu Bruch geworfenen Teile einer Lagerstätte bezeichnet).

Filter pressing

(filtration pressing): The squeezing out of residual magmatic liquids from between crystal masses mashed together by earth movements or other means; a process which in addition to the formation of some types of silicate rocks, is held responsible, by various authors, for the formation of some types of strictly magmatic ore deposits. (Sch.) This term may also be applied to the removal of the connate water during diagenetic crystallization.

Filtro a presión (Filtración a presión): Estrujamiento de líquidos magmáticos residuales entre la masa de cristales triturados conjuntamente por movimientos del terreno o por otros medios; proceso que, además de la formación de algunos tipos de rocas de silicatos, se considera, por muchos autores, como responsable de la formación de algunos tipos de depósitos minerales estrictamente magmáticos. Este término podría aplicarse también al removimiento de agua congenita durante la cristalización diagenética.

Effet de filtre presse: Expression des liquides magmatiques résiduels d'entre des masses cristallines écrasées par les mouvements de terrain ou par tout autre cause; processus, qui, en dehors de la formation de certains types de roches silicatées, est tenu par différents auteurs pour responsable de la formation de gisements strictement magmatiques. Ce terme peut s'appliquer aussi au départ de l'eau connée pendant la cristallisation diagénétique.

Filterpressung, Abpressungsfiltration: Das Ausquetschen von restlichen magmatischen Flüssigkeiten aus durch Erdkrustenbewegung oder ähnliche Einflüsse zusammengepreßten Kristallisationsmassen; ein Vorgang, der, neben der Bildung einiger Arten von Silikatgesteinen, von mehreren Forschern als ausschlaggebend für die Bildung einiger Arten zweifellos magmatischer Erzlagerstätten gehalten wird. Die Bezeichnung kann auch für die Entfernung von Porenwasser während der diagenetischen Kristallisation verwendet werden.

Fishing:

Searching for a piece of drill pipe or drilling equipment broken off from the drilling tools and left in the hole. Many special tools have been designed for catching hold of the missing tool in the well. (Levorsen)

Pesca: Buscar una pieza de la tubería de perforación o del equipo de perforación desprendida del material de trabajo y que se ha quedado en el agujero. Se han diseñado muchas herramientas especiales para coger piezas perdidas en el hueco.

Sauvetage; Repêchage: Recherche d'un morceau de tige ou d'équipement de sondage qui s'est détaché de l'outillage et qui est resté dans le trou. De nombreux instruments ont été élaborés pour se saisir de la partie restée au fond du puits.

Fangarbeit: Die Suche und das Zutagebringen eines Teiles der Bohrausrüstung, der in das Bohrloch gefallen oder darin verblieben ist. Viele Spezialwerkzeuge wurden für solche Zwecke entwickelt.

Fissure (see Figs. 2a, b):

Tabular crack, break or opening in rock. Usually not more than a few meters in width, but sometimes extending to several kilometers (cf. joint fissures and fissure veins).

Fisura: Grieta tabular, rotura o abertura en la roca. Generalmente no tiene más que unos pocos metros de ancho, pero a veces se extiende a muchos kilómetros (ver junturas y vetas de fisuras).

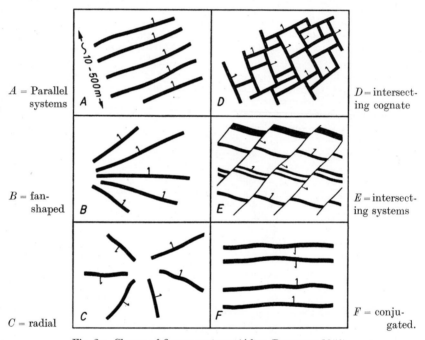

A = Parallel systems

D = intersecting cognate

B = fan-shaped

E = intersecting systems

C = radial

F = conjugated.

Fig. 2a. Classes of fissure systems (After BATEMAN 1956).

Fissure: Discontinuité, cassure ou ouverture tabulaire dans une roche. Habituellement ne dépassant pas quelques mètres en largeur mais pouvant s'étendre sur plusieurs kilomètres (voir diaclase et filon de fissure).

Spalt, Spalte: Flächenartiger Bruch oder Sprung im Gestein. Gewöhnlich nur wenige Meter weit, aber oft auch über mehrere Kilometer sich erstreckend (siehe Absonderungsflächen, Teilungsflächen, Spaltengang).

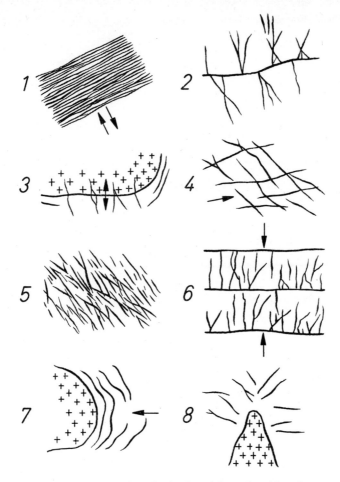

Fig. 2.b Common fissure patterns in veined mineral desposits (After Gromow, 1958).

Fissure vein:

See *vein*.

Vetas de fisuras: Ver veta.
Filon de fissure: Voir filon.
Spaltengang: Siehe Gang.

Flat:

A horizontal deposit especially of ore (in places connected with cross-cutting fracture fillings called *pitches*); also any horizontal, lateral extension of a vein.

Llano: Depósito horizontal, especialmente de mineral (en algunos lugares conectado con fracturas de relleno de corte transversal llamadas *pitches*); también cualquier extensión lateral horizontal de una veta.

Flat: Gîte horizontal, surtout de minerai (par endroits en liaison avec des remplissages de fracture transversale appelés *pitches*); se dit aussi d'une extension latérale horizontale d'une veine.

N.B. – Le terme *flat* est utilisé quelquefois en français pour désigner les surfaces de nivellement alluvionnaire.

Flachung: Eine sich horizontal erstreckende Lagerstätte oder ihr horizontaler Teil (an Stellen mit durchkreuzenden Spaltfüllungen *pitches* oder Einschieben genannt); auch jede horizontale Ausweitung eines Ganges.

Float:

Loose fragments of rock, ore, or gossan found on, or near the surface or in stream beds. (McKinstry)

Rodado: Fragmentos sueltos de roca, mena o sombrero de hierro encontrados cerca de la superficie, en ella misma o en lechos de río.

Roulant: Fragments de roche, de minerai ou de chapeau de fer trouvés près de la surface, ou à la surface, ou dans des lits de rivière.

—: Lose Bruchstücke von Gestein, Erz oder Eisernem Hut, das an, oder nahe der Oberfläche oder in Flußbetten gefunden wird.

Flotation:

A method of concentrating ore by inducing the particles of ore to float to the surface of water or other solution (usually buoyed up by air bubbles) while the gangue particles sink to the bottom. (McKinstry)

Flotación: Método de concentración de mena que induce las partículas de ésta a flotar en la superficie del agua u otra solución (generalmente sostenidas mediante burbujas de aire), mientras que las partículas de la ganga se van al fondo.

Flottation: Méthode de concentration des minerais qui entraine les particules de minerai à flotter à la surface de l'eau ou des solutions (habituellement soutenues par des bulles d'air) alors que les particules de gangue tombent au fond.

Flotation: Eine Methode der Erzkonzentration durch selektive Trennung in Flüssigkeiten. Durch Zugabe von bestimmten Reagenzien für typische Minerale werden diese in Form von kleinen Körnern an der Oberfläche von Luftblasen auf ·schwemmt. Sie werden in konzentrierter Form abgeschöpft, während die Gangart od‹ unerwünschte Komponenten in der Flotationskammer absinken.

Flux

(smelter): Mineral matter (e. g. sandstone or limestone) mixed wıtl smelting ore or concentrates in order to lower the melting point and in order to form a ; roper slag.

Fundente (fundición): Materia mineral (por ejemplo, arenisca o cal za) mezclada con mena fundida o concentrada a fin de bajar el punto de fusión de ést: y formar una escoria apropiada.

Fondant: Produit (p. ex. grès ou calcaire) mélangé à un minerai ou un concentré pour en abaisser la température de fusion et donner une scorie adéquate.

Flußmittel: Mineralstoff (häufig Sandstein oder Kalk), der dem zu schmelzenden Erz beigemengt wird, um dessen Schmelzpunkt zu senken und eine geeignete Schlacke zu bilden.

Fold (see Figs. 3a, 3b):

A bend in strata or in any planar structure (AGI).

1) Anatomy of a fold (S = anticlinal saddle; M = synclinal trough; Sch = limb; A_m = trough axis; A_s = anticlinal axis; E_a = axial plane). 2) vertical fold. 3) overturned fold. 4) recumbent fold. 5) isoclinal fold or zig zag fold. 6) fanned fold or mushroom fold. 7) box fold. 8) fold with fractured and corrugated parts. 9) shear fold. 10) flowage or rheomorphic fold.

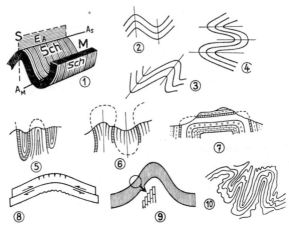

Fig. 3a. Schematic fold patterns (After MURAWSKI, 1963).

Fig. 3b. Diagram showing variety of structures accommodated within the same bulk movement pattern. The derivation (A) shows symmetrical structures and (B) shows assymmetrical or predominantly shear structures. Both give the same bulk effect and all structures as illustrated belong to the same tectonic regime. (From: HARLAND, W.B., and M. B. BAYLY, 1958. Tectonic Regimes, Geol. Mag. XCV, p. 92).

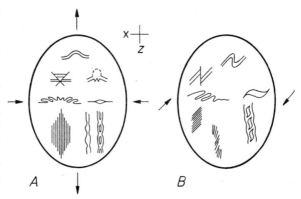

Pliegue: Doblez en una capa o de cualquier estructura plana.

Pli: Inflexion d'une couche ou d'une structure plane quelconque.

Falte: Biegung einer Schicht oder irgendeiner planaren Struktur.

Formation:

(rock-stratigraphic): A formation is the fundamental unit in rock-stratigraphic classification. A formation is a body of rock characterized by lithologic homogeneity; it is prevailingly but not necessarily tabular and may be mapped at the earth's surface or traced in the subsurface. (CSN.)

Formación (litoestratigráfica): Una Formación es la unidad fundamental de la clasificación litoestratigráfica. Una formación es un cuerpo de roca caracterizada por su homogenidad litológica; es predominante pero no necesariamente tabular y puede ser cartografiada en perspectiva plana o ser proseguido en la profundidad.

Formation (lithostratigraphique): Une formation est l'unité fondamentale de la classification lithostratigraphique. Une formation est un ensemble de roches caractérisé par son homogénéité lithologique. Elle est en général, mais non pas nécessairement, tabulaire et peut être cartographiée à la surface de la terre ou suivie en profondeur.

Formation (gesteinsstratigraphisch): Eine Formation ist der fundamentale Begriff in der stratigraphischen Klassifikation. Eine Formation ist ein durch seine lithologische Gleichmäßigkeit gekennzeichneter Gesteinskörper; sie ist vorwiegend, jedoch nicht notwendigerweise flächenhaft; sie kann an der Erdoberfläche oder untertage aufgenommen und kartiert werden.

Fractional crystallization:

Formation of the different crystalline phases of any system (magma, saline waters, charge of a smelter, etc.) at different temperatures and/or partial pressures and concentrations.

Cristalización fraccionada: Formación de diferentes fases cristalinas de cualquier sistema (magma, aguas salobres, carga de una fundición, etc.) a diferentes temperaturas y/o parciales presiones y concentraciones.

Cristallisation fractionnée: Formation des différentes phases cristallines d'un système quelconque (magma, eaux salées, charge d'une fusion), à température différente, et/ou pression partielle et concentration différentes.

Fraktionierte Kristallisation: Die Bildung der verschiedenen Kristallphasen jeglichen Systems (Magma, Salzwasser, Schmelzaufgabe, etc.) bei verschiedenen Temperaturen und/oder Partialdrucken und Konzentrationen.

Fracture:

A break. Fracture is a general term to include any kind of discontinuity in a body of rock if produced by mechanical failure, whether by shear stress or tensile stress. Fractures include faults, shears, joints, and planes of fracture-cleavage. (McKinstry)

Fractura: Rotura. Fractura es un término general que incluye cualquier clase de discontinuidad en un cuerpo de roca si es producido por rotura mecánica, cizallamiento o fuerza de tensión. Las fracturas engloban fallas, cizallamiento, junturas y planos de fractura-exfoliacion.

Fracture: Cassure. N'importe quelle sorte de discontinuité d'une masse rocheuse, si elle est due à une rupture mécanique: cisaillement ou tension. Les fractures englobent failles, diaclases et plans de clivage.

Bruch, Störung: Im allgemeinen ein Ausdruck zur Bezeichnung jeder Art von Diskontinuität in einem Gesteinskörper, die durch mechanischen Bruch, ob Schub- oder

Glory hole:

A large active or inactive mine pit on the surface.

Embudo de extracción (Glory hole): Largo hoyo de mina, activo o inactivo, sobre la superficie.

—: Grande excavation minière à la surface, en activité ou non.

Trichter: Ein ausgedehnter, sich in, oder außer Betrieb befindlicher Tagebau.

Glory hole method:

See *mining methods*.

Método de embudo de extracción: Ver métodos de minería.

—: Voir méthodes minières.

Trichterbau: Siehe Abbaumethoden.

Gossan:

A mixture of oxide material and gangue left on top of sulfide deposits by the surface weathering process. It commonly contains much limonite and sometimes also some trace metal values and boxworks helpful in exploration.

Gosan: Mezcla de óxidos y ganga que quedan en el techo de los depósitos de sulfuros, debido al proceso de alteracion meteórica en la superficie. Generalmente contiene mucha limonita y a veces trazas metálicas y moldes de minerales disueltos que ayudan en la exploración. (sinon. montera de hierro).

Chapeau de fer: Mélange d'oxydes et de gangue laissé au-dessus d'un gisement sulfuré par les processus d'altération superficiels. Il contient généralement beaucoup de limonite et quelquefois des métaux en traces et des *boxworks* utiles pour la prospection.

Gossan, Hut, Eiserner Hut: Ein Gemenge von oxydiertem Material und Ganggestein, das als Rückstand eines Verwitterungsprozesses über einer sulfidischen Lagerstätte verbleibt. Der Rückstand enthält häufig viel Limonit (Eiserner Hut) und öfters auch Spurenmetalle und *boxworks*, die für die Erzprospektion aufschlußreich sind.

Gouge:

Soft fine grained material, often clayey, in faults and fissures. Can be "gouged" with a pick.

Salbanda: Material suave de grano fino, frecuentemente arcilla, situado en fallas y fisuras. Puede moldearse con un pico.

—: Matériel fin, souvent de l'argile, qui remplit des failles et des fissures. Peut être graté au pic. Se traduit en français par salbande lorsque la fissure est minéralisée.

Lettenbelag: Weiches, feinkörniges Material, häufig Ton, auf Störungen und Klüften.

Grade (see Fig. 4):

The metal content or tenor of an ore expressed in percents, ounces per ton, etc.

Ley: Contenido metálico o constitución de una mena expresado en porcentajes, onzas por tonelada, etc.

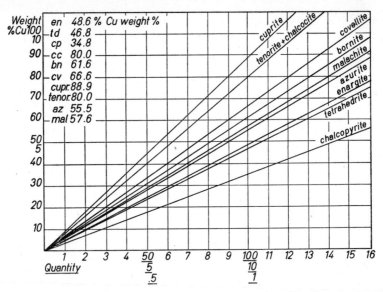

Fig. 4. Chart for relationship between percent copper minerals and percent copper (grade).

Teneur: Contenu en métal d'un minerai, exprimé en %, g/t, etc. . . .

Gehalt: Der Metallinhalt oder -reichtum eines Erzes ausgedrückt in Prozenten oder in Gewichtseinheiten pro Tonne, usw.

Graded bedding (see Appendix V):

Change of grain-size from the bottom to the top of a bed, often with repetitions. Normally the gradation is from coarse at the bottom to fine at the top, with an abrupt change at the bottom of the overlying bed. In fine grained rocks the gradation is sometimes emphasized by change of color from light to dark.

Estratificación gradacional: Cambio de tamaño de los granos desde el fondo al techo, de una capa o sucesión de capas. Normalmente la gradación va de grueso en el fondo a fino en el techo, con un cambio repentino en el fondo de la capa superyacente. En rocas de grano fino la gradación se acentúa a veces por el cambio de color del claro al oscuro.

Granoclassement: Variation de la dimension des grains du haut en bas d'un lit ou d'une succession de lits. Normalement la gradation va des grossiers à la base aux fins au sommet, avec un passage abrupt à la base du lit suivant. Dans des roches à grain fin, la gradation est quelquefois soulignée par un changement de la couleur du clair au sombre.

Gradierte Schichtung: Die Änderung der Korngröße vom Liegenden zum Hangenden einer Sedimentschicht oft mit Wiederholungen. Normalerweise verläuft der Übergang vom Groben am Liegenden zum Feinen am Hangenden mit einer plötzlichen Veränderung gegen das Liegende der überlagernden Schicht. In feinkörnigeren Gesteinen ist die Abstufung noch durch eine Änderung des Farbtons von hell zu dunkel verstärkt.

Gravitational differentiation:

Differentiation accomplished through the gravitational field, as by sinking of a heavy phase (liquid or crystals), or the rising of a light phase (liquid, crystals or gases) through the magma. (AGI.) (cf. *magmatic differentation*).

Diferenciación gravitacional: Diferenciación que tiene lugar a través del campo gravitacional, como el hundimiento de una fase pesada (líquida o cristales), o el ascenso de una fase ligera (líquida, cristales o gases) a través del magma.

Différenciation par gravité: Différenciation due au champ de gravité, telle la descente d'une phase lourde (liquide ou cristaux) ou la montée d'une phase légère (liquide, cristaux ou gaz) à travers le magma.

Gravitationsdifferentiation: Die Differentiation, die durch die Schwerkraft bewirkt wird, z. B. wenn in einem Magma schwere Komponenten absinken (Kristalle usw.) und leichte aufsteigen (Flüssigkeiten, Gase usw.).

Greisen

(K. C. v. LEONHARDT, 1823): The alteration product of a granitic rock, usually in connection with tin mineralization, the new minerals being quartz, muscovite, (lithium-) mica, topaz, tourmaline; the stage of alteration is usually considered to be pneumatolytic.

Greisen (K. C. v. LEONHARDT, 1823): Producto de alteración de una roca granítica, generalmente asociada a una mineralización de estaño en la que los nuevos minerales son cuarzo, moscovita, mica litinífera, topacio, turmalina. El estado de alteración es considerado generalmente neumatolítico.

Greisen (K. C. v. LEONHARDT, 1823): Produit d'altération d'une roche granitique, lié habituellement à une minéralisation stannifère, les minéraux nouveaux étant: quartz, muscovite (lithinifère), mica, topaze, tourmaline. La phase d'altération est généralement considérée comme pneumatolytique.

Greisen (K. C. v. LEONHARDT, 1823): Das Umwandlungsergebnis eines granitischen Gesteins, gewöhnlich in Verbindung mit der Zinn-Mineralisation, wobei als neue Minerale Quarz, Muskovit, (Lithium-) Glimmer, Topas und Turmalin auftreten; die Art der Umwandlung wird als pneumatolytisch bezeichnet.

Ground water:

(ground water = phreatic water). The part of the subsurface water which is in the zone of saturation (A. G. I.). (See vadose water).

Aguas subterráneas: (agua subterránea = agua freática). Parte del agua situada debajo de la superficie, que está contenida en la zona de saturación. (Ver agua vadosa).

Eau souterraine: (eau souterraine = eau phréatique): La part de l'eau souterraine qui se trouve dans la zone de saturation. (Voir eau vadose).

Grundwasser: (Grundwasser = phreatisches Wasser). Der Teil des Wassers unter der Erdoberfläche, welches in der Sättigungszone enthalten ist. (Siehe vadoses Wasser.)

Group

(rock-stratigraphic): A group is the rock-stratigraphic unit next higher in rank than a *formation*; a group consists of two or more associated formations.

Grupo (litoestratigráfica): Grupo es la unidad litoestratigráfica inmediatamente superior a la formación; un grupo contiene dos o más formaciones asociadas.

Groupe lithostratigraphique: Le groupe est l'unité lithostratigraphique immédiatement supérieure à la formation; un groupe contient deux ou plusieurs formations associées.

Gruppe (gesteinsstratigraphisch): Die Gruppe ist die auf die Formation nächstfolgende stratigraphische Einheit nach oben hin; eine Gruppe besteht aus zwei oder mehr zusammengefaßten Formationen.

Gun perforating:

A method of completing producing wells by shooting steel bullets through the casing into the producing formation. The mechanism for firing the bullets is operated electrically, and the effect is to open up the reservoir rock to the well bore at the exact depth for the greatest production, as indicated by various logging and sample data. (LEVORSEN)

Perforación a disparo: Método para completar la ejecución de pozos mediante el disparo de balas de acero a lo largo de la tubería de perforación dentro de la formación productiva. El mecanismo de disparo de las balas es operado eléctricamente y el objeto es hacer accesible la roca reservorio para perforar el pozo a una exacta profundidad y para un máximo de producción, según se conoce por los informes de estudios de testigos de sondaje y muestreo.

Perforation à balles: Méthode de mise en production de sondages consistant à tirer des billes d'acier à travers le tubage dans la formation productive. Le mécanisme de mise à feu des balles est commandé électriquement et l'effet en est d'ouvrir la roche réservoir à la profondeur assurant la plus grande production, telle qu'elle a été indiquée par différents *loggings* et les données d'échantillonnage.

Perforationsschießen: Eine Methode, nach der bei einer Produktionsbohrung die Bohrlochverrohrung im Niveau der Trägerschichten mit Stahlkugeln durchschossen wird. Die Schüsse werden elektrisch gezündet und sie sollen die Lagerstätte gegen die Bohrung hin öffnen und zwar genau in dem Horizont, der nach der Log-Interpretation und nach Probenahmen als der ergiebigste angesehen werden muß.

Hammock structure:

Name for two systems of veins intersecting at an acute angle.

Estructura hamaca: Nombre para dos sistemas de vetas que se intersectan en ángulo agudo.

—: Deux systèmes de filons se recoupant à angle aigu.

—: Bezeichnung für zwei Gangsysteme, die einander in spitzem Winkel schneiden (wie das Gewebe einer Hängematte).

Horse

(in mining): 1. A large block of unmineralized rock included in a vein. 2. Rock occupying a channel cut into a coal bed. 3. Ridge of limestone rising from beneath residual phosphate deposits in Tennessee. 4. In structure: large block of displaced wall rock caught along a fault, particularly a high angle normal fault. (AGI.)

Caballo (en minería): 1. Bloque grande de roca estéril incluida en una veta. 2. Roca que ocupa un canal de corte dentro de un yacimiento de carbón. 3. Loma de caliza que emerge debajo de los depósitos residuales de fosfatos en Tennessee. 4. Bloque grande de roca encajante desplazada y atrapada a lo largo de la falla.

Nerf (en terminologie minière): 1 – Grand bloc stérile enfermé dans une veine. 2 – Roche occupant un chenal taillé dans une couche de charbon. 3 – Arête de calcaire s'élevant de dessous des gisements résiduels de phosphate dans le Tennessee. 4 – En géologie structurale, grand bloc de roche encaissante pris le long d'une faille et plus particulièrement d'une faille normale très inclinée. N.B. Plus généralement le terme nerf désigne en français une passée stérile dans une formation minéralisée.

—: Eine im Bergbau übliche Bezeichnung für 1. einen großen Block oder eine Scholle nicht mineralisierten Gesteins, der in einem Gang eingelagert ist, 2. Gesteinsmassen, die in der Kohle ausgewaschene Kanäle ausfüllen, 3. Kalkrücken in Tennessee, die sich über tiefer liegende Phosphatlagerstätten erheben, und 4. einen großen Block ausgebrochenen Nebengesteins, der in einer Verwerfungszone eingeschlossen ist, besonders in steilstehenden Verwerfungen.

Horsetail structure:

An arrangement in the shape of a horsetail, of closely spaced mineralized fissures branching out from major veins and forming together a stockwork, (Sch.) (typical examples in Butte, Montana, U.S.A.).

Estructura de cola de caballo: Disposicion en forma de cola de caballo de fisuras mineralizadas y estrechamente separadas, las cuales se ramifican de las vetas principales y forman juntas un stockwork, ejemplos típicos en Butte, Montana, EE.UU.).

Structure en queue de cheval: Arrangement en forme de queue de cheval de fissures minéralisées très rapprochées se séparant d'une veine principale et formant un stockwork, (exemples typiques à Butte, Montana, U.S.A.).

Pferdeschwanzstruktur: Eine Erzanordnung in Form eines Pferdeschwanzes, bestehend aus nahe aneinander liegenden mineralisierten Feinbrüchen, die von stärker ausgebildeten Gängen abzweigen und zusammen ein Stockwerk bilden, (typische Beispiele in Butte, Montana, USA).

Hybrid rock or ore:

Formed by admixture of other rock or ore matter (used mostly for magmatic rocks).

Roca o mena híbrida: Formada por la incorporación de otro cuerpo de roca o mena (usado generalmente para rocas magmáticas).

Roche (ou minerai) hybride: Formé par addition d'une autre roche ou de matière minérale. (Terme pue utilisé en français.)

Hybrides Gestein oder Erz: Gebildet durch Vermischung mit anderem Gestein oder Erzmaterial (meist für magmatische Gesteine verwendet).

Hydraulic mining:

See mining methods.

Minería hidráulica: Ver métodos de minería.

Abattage hydraulique: Voir méthodes minières.

Hydraulische Gewinnung: Siehe Abbaumethoden.

Hydrothermal (see Fig. 18 and Appendix VIIa, b; VIIIa, b, c):

Material or processes which include water at higher than room temperature, up to about 400 °C (about 800 °F), originating from igneous or metamorphic rocks (cf. alteration and deuteric). (See Appendix Fig. VII and VIII).

Hidrotermal: Material o procesos que incluyen agua a temperatura más alta que la del medio ambiente, pudiendo llegar hasta cerca de 400 °C y que se origine en rocas ígneas o metamórficas (Ver alteración y deutérico).

Hydrothermal: Matériel ou processus comportant de l'eau à des températures supérieures à la température ambiante et pouvant monter jusqu'à 400 °C, tirant son origine de roches ignées ou métamorphiques. (Voir altération et deutérique).

Hydrothermal: Vorgänge, Mineralbildungen an denen Wasser mit einer Temperatur höher als Raumtemperatur bis zu etwa 400 °C beteiligt ist; ist in Zusammenhang mit Erstarrungsgesteinen und metamorphen Gesteinen zu sehen, (siehe Umwandlung und deuterisch).

Hypabyssal:

A general term applied to minor shallow intrusions, such as sills and dikes, and to rocks of which they are made. They are distinguished from volcanic rocks and formations on the one hand and "plutonic" rocks and major intrusions, such as batholiths, on the other.

Hipoabisal: Término general aplicado a intrusiones menores y superficiales, tales como sills (filones capas) o diques y a rocas que se forman de ellas. Se diferencian por un lado de las rocas y formaciones volcánicas y por otro de las rocas plutónicas e intrusiones mayores.

Hypoabyssique: Terme général appliqué aux intrusions mineures de faible profondeur, telles que sills et dikes et aux roches qui les constituent. On les distingue des roches et formations volcaniques d'une part et des roches et intrusions majeures «plutoniques», telles que les batholites, d'autre part.

Hypabyssisch: Ein allgemeiner Ausdruck für weniger umfangreiche, oberflächennahe Intrusionen, wie z. B. für Lagergänge und Kluftfüllungen oder Gesteine, aus denen diese bestehen. Sie werden unterschieden von Vulkangesteinen und Formationen einerseits, und von Tiefengesteinen und größeren Intrusionen wie Batholithen andererseits.

Hypogene:

Generated from depth. Refers to the effects produced by ascending (usually hydrothermal) solutions. (Cf. *Supergene.*) (MCKINSTRY)

Hipógeno: Generados en la profundidad. Se refiere a los efectos producidos por soluciones ascendentes (generalmente hidrotermales). (Ver supergeno).

Hypogène: Né de la profondeur. Désigne en général les effets produits par des solutions ascendantes (habituellement hydrothermales). (Voir supergène).

Hypogen: Aus der Tiefe stammend; bezieht sich auf Auswirkungen, die von aufsteigenden (gewöhnlich hydrothermalen) Lösungen stammen (siehe supergen).

Impregnation:

Designation for a mineral deposit with a diffuse or disseminated arrangement of the ore minerals (also: the process of formation of such an arrangement, but without any *a priori* connotation as to time, source and cause).

Impregnación: Denominación de un depósito mineral con una distribución de menas esparcidas o diseminadas (Designa también el proceso de formación de tal distribución, pero sin ninguna connotación *a priori*, como tiempo, origen y causa).

Imprégnation: Désigne un gîte minéral avec une répartition diffuse ou disséminée des minerais. (Désigne également le processus de formation d'une telle répartition sans implications *a priori* quant à l'époque, la source et la cause).

Imprägnation: Bezeichnung einer Erzlagerstätte mit einer fein verteilten oder *disseminated* Anordnung der Minerale. (Auch: der Vorgang der Bildung einer solchen Anordnung, aber ohne vorherige Bestimmung der Relationen von Zeit, Ursprung und Ursache).

Infra-, intracrustal:

Formed within the crust of the earth (these terms to be used in place of *endogenous*, cf.). (opp.: *supracrustal*.)

Infra-, intracortical: Formada dentro de la corteza de la tierra (estos términos pueden usarse en lugar de endógeno, véase esta voz). (contrario: supracortical.)

Infra- et intracrustal: Formé à l'intérieur de l'écorce terrestre. (Termes à utiliser à la place d'endogène, voir ce mot.) (contraire: supracrustal.)

Infra-, intrakrustal: Innerhalb der Erdkruste gebildet (diese Ausdrücke können auch anstelle von endogen verwendet werden). (Gegenteil: suprakrustal.)

Intramagmatic (see Appendix VIIb):

Deposits: formed within the original magma. Stage: main crystallization period of a magma.

Intramagmático: Depósitos formados dentro del magma original. Fase principal de cristalización de un magma.

Intramagmatique: Gîte: formé à l'intérieur du magma original. Phase: phase de cristallisation principale d'un magma.

Intramagmatisch: Intramagmatische Lagerstätten: innerhalb des ursprünglichen Magmas gebildet. Intramagmatisches Stadium: Hauptkristallisationsperiode eines Magmas.

Joint:

A divisional plane or surface in the form of a smooth fracture, that divides a rock and along which there has been no visible movement parallel to the plane or surface. (BILLINGS)

Diaclasa, juntura: Plano o superficie divisoria en forma de fractura sencilla que divide una roca y a lo largo del cual no ha habido movimiento visible paralelo al plano o a la superficie.

Diaclase: Plan ou surface de division en forme d'une fracture simple d'une roche le long duquel ou de laquelle il n'y a pas eu de mouvement parallèle au plan ou à la surface.

Kluft, Absonderungsfläche: Einfache Bruchebenen im Gestein, die eine deutliche Trennung erkennen lassen, ohne daß eine Verschiebung beider Seiten gegeneinander entlang der Ebene sichtbar festgestellt werden kann.

Juvenile:

Coming to the surface for the first time; fresh, new in origin; applied chiefly to gases and waters. (AGI.)

Juvenil: Que viene a la superficie por primera vez; aplicado principalmente a gases y aguas.

Juvénile: Venant pour la première fois à la surface; frais, d'origine nouvelle; s'applique surtout aux gaz et aux eaux.

Juvenil: Zum erstenmal an die Oberfläche austretend; frisch, in der Entstehung; meist für Gase und Wasser verwendet.

Katathermal ore deposits (see Appendix VIIb):

Hydrothermal deposits formed at high temperatures.

Depósitos minerales catatermales: Depósitos hidrotermales formados a alta temperatura.

Gîte catathermal: Gîte hydrothermal formé à haute température.

Katathermale Erzlagerstätten: Bei hohen Temperaturen gebildete hydrothermale Lagerstätten.

Kryptomagmatic deposits (see Appendix VIIb):

Formed outside of, and in hidden connection with a parent intrusion.

Depósitos criptomagmáticos: Formados fuera de la intrusión generatriz y en conexión oculta con ella.

Gîte cryptomagmatique: Gîte formé à l'extérieur de l'intrusion qui lui a donné naissance et dont la relation avec cette intrusion est cachée.

Kryptomagmatische Lagerstätten: Außerhalb von und in versteckter Verbindung zu einer Intrusion gebildete Lagerstätten.

Laccolith:

Lens-shaped mass of igneous rock intrusive into layered rocks. Typically a laccolith has a flat floor and a domed roof and is more or less circular in ground plan (Long-well)

Lacolito: Masa de roca ígnea intrusiva en forma de lente. Un lacolito típico tiene una base plana y un techo on forma de cúpula y tiene una proyección horizontal más o menos circular.

Laccolite: Masse lenticulaire de roche ignée intrusive dans des roches stratifiées. Un laccolithe a typiquement une base plane et un sommet en coupole et a une projection horizontale plus ou moins circulaire.

44

Lakkolith: Eine linsenförmige Masse aus Tiefengestein, die in geschichtetes Gebirge eingedrungen ist. Kennzeichnenderweise hat ein Lakkolith eine flache untere Begrenzungsebene, eine domartige nach oben und meist einen kreisförmigen Grundriß.

Ladder vein:

See *vein*.

Veta escalonada: Ver veta.

Filon en échelle: Voir filon.

Leitergang: Siehe Gang.

Lateral secretion theory

(SANDBERGER, 1877): The theory that the contents of a vein or lode are derived from the adjacent country rock by a leaching process, whereby either superficial water or thermal water is involved. (SCH.)

Teoría de secreción lateral (SANDBERGER, 1877): Teoría que sostiene que el contenido de una veta o filón se ha derivado de la roca de caja adyacente mediante un proceso de lixiviación de aguas superficiales o termales.

Latéralisme (SANDBERGER, 1877): Théorie admettant que le contenu d'une caisse filonienne ou d'un lode est dérivé de la roche encaissante adjacente par un processus de lessivage par des eaux superficielles ou thermales.

Lateralsekretion (SANDBERGER, 1877): Die Theorie, nach der eine Gangfüllung durch einen Auslaugungsprozeß aus dem umgebenden Nebengebirge stammt, wobei entweder Oberflächen- oder Thermalwasser mitgewirkt hat.

Lenticular vein:

See *vein*.

Veta lenticular: Ver veta.

Filon lenticulaire: Voir filon.

Linsengang: Siehe Gang.

Level

(in a mine): Group of workings all at approximately the same elevation. In most mines levels are spaced at regular intervals of depth, usually 100 to 200 feet apart. (MCKINSTRY)

Nivel (en una mina): Grupo de trabajos situados aproximadamente a la misma elevación. En la mayoría de las minas los niveles están separados por intervalos regulares de profundidad, generalmente de 100 a 200 pies.

Niveau (dans une mine): Ensemble de travaux miniers situés approximativement à la même altitude. Dans la plupart des mines, les niveaux sont espacés régulièrement et séparés par des tranches de 30 à 65 m.

Sohle (eines Bergwerks): 1. Ein System von Grubenbauen, die sich in derselben Horizontalebene befinden. In den meisten Bergwerken sind die Sohlen in regelmäßigen Abständen angeordnet, die 30 bis 65 m betragen. 2. Der „Boden" jeden bergmännischen Hohlraumes.

Linked veins:

See *vein*.

Vetas eslabonadas: Ver veta.

Filons anastomosés: Voir filon.

Kettengang: Siehe Gang.

Liquid magmatic (see Appendix VIIb):

Name of the earliest stage of magmatic crystallization and differentiation, when the magma is still essentially complete and a "hot silica gel".

Líquido magmático: Nombre del primer estado de cristalización o diferenciación magmática, cuando el magma está todavia esencialmente completo y en un estado de "sílice gelatinosa caliente".

Protocristallisation: Phase la plus précoce de la cristallisation et différenciation magmatique, quand le magma est encore essentiellement complet et est encore un «bain silicaté chaud».

Liquidmagmatisch: Die Bezeichnung für das früheste Stadium der magmatischen Kristallisation und Differentiation, in dem das Magma noch eine vollständige, d. h. heiße Silikatschmelze ist.

Lode:

(s. s.); composite vein; composite lode: A large fractured zone consisting of a number of approximately parallel fissures filled with ore, and occasionally also of partially replaced country rock. (SCH.)

Mother lode: The original lode deposit from which a placer deposit has been derived, or the principal lode or vein passing through a particular district or section of the country.

Venero: (Veta compuesta, filón compuesto); Zona fracturada grande, que consiste en un número de fisuras aproximadamente paralelas y rellenadas con mineral y ocasionalmente, la roca encajante parcialmente reemplazada.

Criadero: Depósito original del cual se ha derivado un placer. Además: veta principal que pasa a través de un distrito particular o sección de un terreno.

—: —strictement parlé: filon complexe. — lode complexe: grande zone fracturée comportant nombre de fissures approximativement parallèles remplies de minerai et éventuellement aussi de roche encaissante partiellement remplacée.

—: Gîte de lode primaire duquel un placer a été dérivé. Encore: lode ou filon principal traversant un district ou une zone particulière à une région.

—: Ein zusammengesetzter Gang; eine große bruchreiche Zone aus einer Anzahl ungefähr parallel laufender Bruchspalten, die mit Erz und auch gelegentlich mit teilweise verdrängtem Nebengestein gefüllt sind.

Mother Lode (engl.): Eine Ursprungslagerstätte, von der das Erz gelöst und in Seifen-form zu einer anderen Stelle bewegt worden ist; auch der Hauptgang, der durch eine erzreiche Gegend streicht.

Longwall:

See *mining methods*.

Longwall: Ver métodos de minería.

Longue taille: Voir méthodes minières.

Streb: Siehe Abbaumethoden.

Lopolith:

A concordant intrusion associated with a structural basin. In the simplest and ideal case the sediments above and below the lopolith dip inward toward a common center (BILLINGS)

Lopolito: Intrusión concordante asociada con una cubeta estructural. En el caso más simple e ideal los sedimentos por encima y por debajo del lopolito buzan hacia dentro y hacia un centro común.

Lopolite: Intrusion concordante associée à un bassin structural. Dans le cas idéal le plus simple, les sédiments au-dessus et en-dessous du lopolite plongent vers l'inté-rieur en direction d'un centre commun.

Lopolith: Eine konkordante Intrusion in einem geologischen Becken. Im einfachsten Falle tauchen die Schichten oberhalb und unterhalb des Lopoliths zur Mitte gegen ein gemeinsames Zentrum ein.

Magma:

Molten rock matter.

Magma: Materia de roca fundida.

Magma: Matériau rocheux fondu.

Magma: Gesteinsschmelze, wie sie im Erdinneren vor oder nach der Differentiation auftritt.

Magmatic differentiation (see Fig. 5 and Appendix VIIa, VIIIa, b, c):

The process of separation into fractions of different composition and textures in a molten igneous mass (magma). It takes place in the liquidmagmatic stage by gravi-tative crystal settling (crystallisation differentiation); in the pegmatitic or pneumato-lytic and in the hydrothermal stage by mere fractional crystallisation. The liquid and/or volatile fraction may remain or escape from the solidified mass.

Diferenciación magmática: Proceso de separación en fracciones de diferente composi-ción y texturas, en una masa ígnea fundida (magma). Ello se produce en el estado líquido-magmático por depositación gravitacional de cristales (diferenciación por cristalización); en el estado pegmatítico o neumatolítico y en el hidrotermal, más bien por cristalización fraccionada. La fracción líquida o volátil puede quedar en la masa solidificada o escapar de ella.

47

Différenciation magmatique: Processus de séparation en fractions de composition et de textures différentes dans une masse ignée fondue (magma). Cell-ci à lieu à l'état liquide-magmatique par ségrégation des cristaux par gravité (différenciation par cristallisation); à l'état pegmatitique ou pneumatolytique comme à l'état hydrothermal la séparation se fait par cristallisation fractionnée. La fraction liquide et/ou volatile peut rester ou s'échapper de la masse solidifiée.

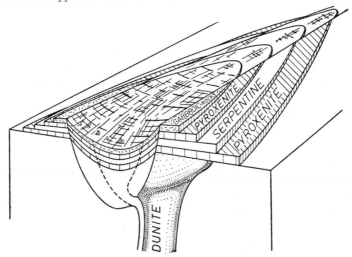

Fig. 5. Type figure of ore deposit formed by differentiation crystallization (crystal settling) in an ultrabasic magma. (Example: Great Dyke of Rhodesia. From WORST, 1958, p. 283).

Magmatische Differentiation: Der Vorgang der Trennung in einzelne Fraktionen individueller Zusammensetzung und Struktur durch Auskristallisieren einer natürlichen Silikatschmelze (Magma). Er besteht aus verschiedenen Stadien, die im wesentlichen in drei Teile zerfallen (entsprechend dem Differentiationsschema der Figuren in Appendix VIIa, VIIIa, b, c); im liquidmagmatischen Teil im wesentlichen in gravitative Kristallabsaigerung, im pegmatitisch-pneumatolytischen und im hydrothermalen Teil in Trennung in kristallisierte und noch flüssige oder leichtflüchtige Teile.

Main vein:
See *vein*.

Veta principal: Ver veta.
Filon principal: Voir filon.
Hauptgang: Siehe Gang.

Manto:
A blanket-like ore body, usually confined to a stratigraphic horizon. (In some Latin American countries also any flattish pipe or stratum or irregular flat mass of ore.)

Manto: Yacimiento mineral horizontal de forma tabular, generalmente restringido a un horizonte estratigráfico. (En algunos países latinoamericanos, también cualquier depósito achatado, estrato o masa irregular plana de mineral).

Manto: Terme espagnol. Corps minéralisé de forme stratoïde habituellement limité à un horizon stratigraphique. (Dans quelques pays d'Amérique latine désigne également toute colonne ou couche plate ou toute masse irrégulière, mais aplatie, de minerai.)

Manto: Ein großflächiger Erzkörper, der gewöhnlich an einen stratigraphischen Horizont gebunden ist. (In einigen Ländern Latein-Amerikas auch flache Schlote oder Schichten oder jede irreguläre flache Erzmasse.)

Matte:

A product of smelting, richest in metal but still containing some sulfur; goes to the blast furnace or refinery for further treatment and purification.

Mata: Producto de fundición rico en metal, pero conteniendo todavía azufre. Se trata en el horno de fundición o refinería para un tratamiento y purificación adicional.

Matte: Produit de fusion, très riche en métal, mais contenant toujours du soufre. Va au convertisseur ou à la raffinerie pour traitement et purification ultérieurs.

Matte: Ein Verhüttungsprodukt mit sehr angereichertem Metall, doch noch etwas schwefelhaltig; es wird zur weiteren Veredlung im Schachtofen oder in einer Raffinieranlage behandelt.

Member:

(rock-stratigraphic): A member is a part of a formation; it is not defined by specified shape or extent. A geographically restricted member that terminates on all sides within a formation may be called a *lentil*. A member that extends outward beyond the main body of a formation may be called a *tongue*. (CSN.)

Miembro: (término estratigráfico) Miembro es una parte de una formación y no está definido por una forma o extensión específica. Un miembro geográficamente restringido, cuyos lados terminan todos dentro de una formación, puede llamarse *lenteja*. Un miembro que se extiende más allá del cuerpo principal de una formación puede llamarse una *lengua*.

Membre lithostratigraphique: Un membre est une partie d'une formation; il n'est pas défini par une forme ou une étendue déterminées. Un membre géographiquement limité, qui se termine de tous les côtés à l'intérieur d'une formation, peut être appelé lentille. Un membre qui s'étend au-dehors et au-delà d'une formation principale peut être appelé *langue*.

Schichtglied (stratigraphisch): Ein Teil einer Schicht oder Formation, nicht genauer bestimmt durch typische Formen oder Ausdehnungen. Ein geographisch begrenztes Schichtglied, das in allen Richtungen in einer Formation begrenzt ist, kann Linse, ein Glied, das aus dem Hauptkörper einer Formation hinausreicht, kann *Zunge* genannt werden.

Merismitic fabric:

A fabric characterized by the irregular interpenetration of its units. (App. VIb).

Estructura merismática: Estructura caracterizada por la interpenetración irregular de sus unidades. (ver App. VIb).

—: Structure caractérisée par une interpénétration irrégulière de ses éléments. (Voir App. VIb).

Merismitisches Gefüge: Eine geometrische Anordnung, die durch das irreguläre gegenseitige Durchsetzen ihrer einzelnen Bestandteile gekennzeichnet ist. (Siehe App. VIb).

Mesothermal ore deposit (see Appendix VIIb):

Deposit formed by hydrothermal solutions at intermediate temperature.

Depósito mineral mesotermal: Yacimiento formado por soluciones hidrotermales a temperaturas intermedias.

Gîte mésothermal: Gîte formé par des solutions hydrothermales à température intermédiaire.

Mesothermale Lagerstätten: Bei mäßigen (mittleren) Temperaturen gebildete hydrothermale Lagerstätten.

Metallogenic epoch:

A geological span of time during which above average rates or amounts of mineral deposition took place.

Epoca metalogénica: Espacio geológico de tiempo durante el cual tuvo lugar un elevato promedio o cantidad de deposición de mineral.

Epoque métallogénique: Laps de temps géologique pendant lequel des minéralisations se sont produites à une vitesse ou en quantité supérieure à la normale.

Metallogenetische Epoche: Geologische Zeitspanne, in denen überdurchschnittliche Anreicherungen von Erzmineralien stattgefunden haben.

Metallogenic province:

A geographical or geological region which displays more or less uniform and above average quantities of mineral deposits, normally of the same type.

Provincia metalogénica: Región geográfica o geológica que contiene una cantidad mayor o menor, pero superior a la normal, de depósitos minerales, con frecuencia del mismo tipo.

Province métallogénique: Région géographique ou géologique possédant des gîtes minéraux, de type plus ou moins uniforme et en quantité supérieure à la normale, souvent du même type.

Metallogenetische Provinzen: Geographische oder geologische Regionen, in denen ähnliche und ungewöhnlich zahlreiche Mineralvorkommen auftreten, die meist zur gleichen Art gehören.

Metamorphism (see Figs. 6a, b):

The mineralogical and structural adjustment of solid rocks to physical or chemical conditions which have been imposed at depths below the surface zones of weathering and cementation and which differ from the conditions under which the rocks in question originate (TURNER and VERHOOGEN). Cf. *contact metamorphism* (fig. 6a) and *regional metamorphism* (fig. 6b).

50

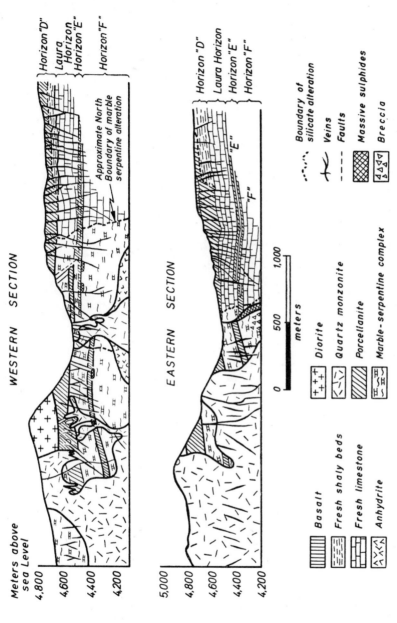

Fig. 6a. Type figure for a contact metamorphic area with contact mantos and vein deposits (Morococha, North-South-sections, after Terrones).

WESTERN SECTION

EASTERN SECTION

Meters above sea Level

Horizon "D"
Laura Horizon
Horizon "E"
Horizon "F"

Approximate North Boundary of marble serpentine alteration

Diorite
Quartz monzonite
Porcellanite
Marble-serpentine complex

Basalt
Fresh shaly beds
Fresh limestone
Anhydrite

Boundary of silicate alteration
Veins
Faults
Massive sulphides
Breccia

500   1,000
meters

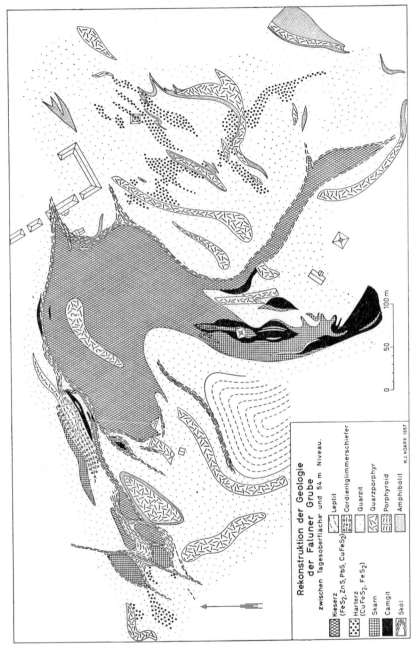

Rekonstruktion der Geologie der Faluner Grube
zwischen Tagesoberfläche und 54 m Niveau.

Kieserz
(FeS₂, ZnS, PbS, CuFeS₂)

Harterz
(CuFeS₂, FeS₂)

Skarn

Camgiit

Sköl

Leptit

Cordieritglimmerschiefer

Quarzit

Quarzporphyr

Porphyroid

Amphibolit

H.J.KOARK 1957

Fig. 6b. Type drawing for a metamorphosed stratiform deposit (the Falun deposit, Sweden; after KOARK, 1957).

Metamorfismo: Ajuste mineralógico y estructural de rocas sólidas bajo condiciones físicas o químicas que les han sido impuestas a profundidades superiores a las zonas de cementación y de alteración superficiales y que son diferentes de las condiciones bajo las que se han formado las rocas consideradas. (véase fig. 6 a y 6 b, y metamorfismo de contacto y metamorfismo regional).

Métamorphisme: Ajustement minéralogique et structural de roches solides à des conditions physiques ou chimiques qui leur ont été imposées à des profondeurs supérieures aux zones de cémentation et d'altération superficielles et qui sont différentes des conditions sous lesquelles les roches considérées se sont formées. Voir métamorphisme de contact (fig. 6 a) et régional (fig.6 b).

Metamorphose: Die mineralogische und strukturelle (chemische und physikalische) Umwandlung fester Gesteine als Reaktion zur Anpassung an physikalische oder chemische Bedingungen, die in der Tiefe außerhalb des atmosphärischen Einflusses auftreten und die sich von den ursprünglichen Entstehungsbedingungen des jeweiligen Gesteins unterscheiden. Siehe Kontaktmetamorphose (Fig. 6 a) und Regionalmetamorphose (Fig. 6 b).

Metasomatism:

Metamorphism involving introduction and removal of certain substances, with corresponding changes not only in the mineralogy but in the chemical composition of the rocks affected (TURNER and VERHOOGEN). C. f. *contact metasomatism.*

Metasomatismo: Metamorfismo que entraña introducción o remoción de ciertas sustancias con los cambios correspondientes no sólo en la composición mineralógica, sino en la química, de las rocas afectadas. (véase: metasomatismo de contacto.)

Métasomatose: Métamorphisme comportant apport et départ de certaines substances avec des modifications correspondantes, non seulement dans la minéralogie, mais encore dans la composition chimique des roches intéressées. Voir métasomatose de contact.

Metasomatose: Eine Metamorphose, bei der Stoffe zu- und abgeführt werden und entsprechende Änderungen mineralogischer und chemischer Natur auftreten. Siehe Kontaktmetasomatose.

Middlings:

Ore middling products; intermediate product of ore dressing, containing mostly gangue plus ore minerals locked to gangue; locked ore minerals sometimes are called "middlings" or "middling particles".

Middlings: La molienda intermedia; producto intermedio en la preparación mecánica de la mena que contiene principalmente gangas más mena trabada en la ganga; las menas trabadas se llaman a veces *middlings* o *middling particles.*

Mixtes: Produits intermédiaires du traitement des minerais, contenant surtout de la gangue et des minéraux utiles encore liés aux grains de gangue. Les particules de minerai enchâssées dans la gangue sont quelquefois appelées «mixtes».

Zwischenprodukte: Ein unfertiges Produkt aus einem Teilprozeß der Aufbereitung, das meist noch Ganggestein und Erz − letzteres an das Ganggestein gebunden − enthält. Solche Teilprodukte mit eingeschlossenen Erzmineralien werden im amerikanischen Sprachraum häufig *middlings* oder *middling particles* genannt.

Milling ore:

A term most frequently used to designate that part of a mineral deposit from which the ore can be mixed with less valuable material, processed in a mill and sold at a profit.

Mineral de molienda: Término usado frecuentemente para designar la parte de un depósito mineral de la cual se puede obtener mineral útil que puede ser tratado en una fábrica de concentración y vendido con provecho.

—: Terme utilisé le plus fréquemment pour désigner la partie d'un gisement dont le minerai peut être mélangé avec des matériaux de moindre valeur, broyé et vendu avec profit.

Reicherz, bauwürdiges Erz: Eine Bezeichnung für jenen Teil von Erzlagerstätten, dessen wertvolles Material mit minderwertigem vermengt, in Mühlen durchgearbeitet und noch mit Profit verkauft werden kann.

Mimetic (inherited) structure:

An original structural feature of the country rock, which has been preserved at least in part after its replacement by ore, or after metamorphic recrystallization.

Estructura mimética (heredada): Característica estructural original de la roca de caja, la cual ha sido preservada, al menos en parte, después de su reemplazamiento por mena, o después de una recristalización metamórfica.

Structure relique: Caractère structural primaire de la roche encaissante, qui a été conservé, au moins partiellement, après son remplacement par le minerai ou après la recristallisation métamorphique.

Mimetische (übernommene) Struktur: Eine Strukturform von Nebengestein, die nach einer Verdrängung durch Erz oder auch nach einer Metamorphose oder Neukristallisation zumindest teilweise erhalten geblieben ist.

Mineral deposit:

Any mass of minerals or any rock, hard or soft, consolidated or non-consolidated, which may be used sooner or later for the recovery of an economic mineral or metal.

Depósito mineral: Cualquier masa de mineral o roca, dura o blanda, consolidada o inconsolidada, que puede ser usada tarde o temprano para la recuperación de un mineral o metal económico.

Gîte minéral: N'importe quelle masse de minéraux, ou n'importe quelle roche, dure ou tendre, consolidée ou non consolidée, qui peut être utilisée tôt ou tard pour l'obtention d'un minéral ou d'un métal de valeur économique.

Minerallagerstätte: Jede Masse eines harten oder weichen, verfestigten oder unverfestigten Gesteins mit Erzgehalt, aus dem früher oder später Mineralien oder Metalle wirtschaftlich gewonnen werden können.

Mineralization:

1. The process of replacing and/or filling the organic constituents of a body by inorganic fossilization. 2. In connection with ore deposits: the presence of ore minerals or the

process of their formation, with no connotation as to when and how it took place, and to whether any material was introduced or not.

Mineralización: 1. Proceso de reemplazar y/o rellenar los constituyentes orgánicos de un cuerpo por fosilización inorgánica. 2. En relación con depósitos minerales: la presencia de menas o el proceso de su formación, sin connotación de cuándo y cómo tuvo lugar o de si hubo o no introducción de mineral.

Minéralisation: 1 – Processus de remplacement et (ou) de remplissage des constituants organiques d'un corps par fosilisation inorganique. 2 – En relation avec les gîtes minéraux, la présence de minerais ou leur processus de formation, sans implication quant à la date ou le mode de mise en place ou à la présence ou l'absence de matière introduite.

Mineralisation: 1. Der Vorgang der Verdrängung und/oder Einlagerung von anorganischen Stoffen in organisches Ursprungsmaterial. 2. Bei Erzlagerstätten: DieGegenwart von Mineralien oder auch der Vorgang ihrer Bildung, ohne eine Bestimmung über die Zeit und Art des Vorgangs und die Zuführung von weiteren Mineralien.

Mineralized water:

Often used for groundwater or spring water containing more than 1 g/l of dissolved mineral matter, but that is of no special composition and use (see *groundwater*, *mineral spring*, *mineral water*, etc.).

Agua mineralizada: Término usado a menudo para designar aguas subterráneas o manantiales que contienen más de 1 g/l de sustancia mineral disuelta, pero que no es de composición y uso especial (ver aguas subterráneas, fuentes de aguas minerales, aguas minerales, etc.).

Eau minéralisée: Souvent utilisé pour désigner l'eau souterraine ou l'eau de source contenant plus d'un g/l de matière minérale dissoute, mais qui n'est ni d'un emploi, ni d'une composition particulière. (Voir source minérale et eau minérale).

Mineralhaltiges Wasser: Grund- oder Quellwasser mit mehr als 1 g/l gelöster Mineralsubstanz, aber oft ohne besondere Zusammensetzung und besondere Verwendung (vgl. Grundwasser, Mineralquelle, Mineralwasser, usw.).

Mineral spring:

A spring whose water is known to contain a certain amount (usually above 1 g/l) of mineral salts and occasionally also gases. The salts being of uncommon composition and occurence, and usually of balneological or therapeutic value.

Fuente mineral: Fuente cuya agua se sabe que contiene una cierta cantidad (generalmente superior a 1 g/l) de sales minerales y a veces algo de gases. Las sales son de composición y ocurrencia poco común y generalmente de valor balneariológico o terapéutico.

Source minérale: Source dont on sait que l'eau contient une certaine quantité (en général supérieure à 1 g/l) de sels minéraux et éventuellement de gaz. La présence et la composition des sels étant inhabituelles et généralement de valeur balnéothérapeutique.

Mineralquelle: Eine Quelle, deren Wasser einen bestimmten Gehalt (gewöhnlich > 1 g/l) an Mineralsalzen und manchmal auch Gasen führt. Die Salze sind meist von ungewöhnlicher Zusammensetzung und von balneologischem oder therapeutischem Wert.

Mineral water:

Groundwater or spring water with a certain quantity (usually above 1 g/l) of mineral salts or gases, being of uncommon composition and occurrence. Often used for balneological or medicinal purposes.

Agua mineral: Agua subterránea con una cierta cantidad (generalmente superior a 1 g/l) de sales o gases minerales que son de composición y ocurrencia poco común. Frecuentemente usada para aplicaciones balneariológicas o medicinales.

Eau minérale: Eau souterraine ou eau de source ayant une certaine quantité (en général plus d'un g/l) de sels minéraux ou gaz, de présence et de composition inhabituelles. Souvent utilisé pour des usages balnéologiques ou médicinaux.

Mineralwasser: Häufig für Grundwasser oder Quellwasser gebraucht, das mehr als 1 g/l Mineralgehalt gelöst enthält, wobei die Zusammensetzung jedoch nicht genau bestimmt ist und auch die Verwendungsmöglichkeiten vielfacher Art sein können, z. B für Heilzwecke (zum Baden oder Trinken).

Mining methods (common examples, mostly after McKINSTRY) (see Fig. 7a, 7b, ...):

Caving: A method of mining in which the ore, the support of a great block being removed, is allowed to cave or fall, and in falling is broken sufficiently to be handled; the overlying strata subside as the ore is withdrawn.

Variations: *Top slicing:* A horizontal slice of ore is removed, allowing the slice above it to cave. Then, leaving the slice below temporarily intact, a still lower slice is removed,

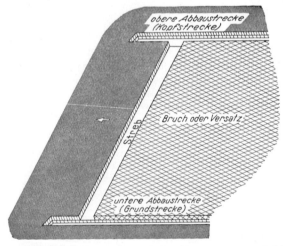

Fig. 7. Mining Methods (all figures after DORSTEWITZ et al. 1959).
   a) Longwall mining in flat-lying ore-bodies

for:		read:	
obere Abbaustrecke		upper mining level	
untere Abbaustrecke		lower mining level	
Strebraum		working place	
Bruch oder Versatz („Alter Mann")		fill (waste)	

allowing the intervening slice to cave. The process is repeated until the bottom of the orebody is reached. *Block caving:* Similar to top slicing except that the slice which is allowed to cave is of much greater thickness and may, in fact, constitute the full thick-

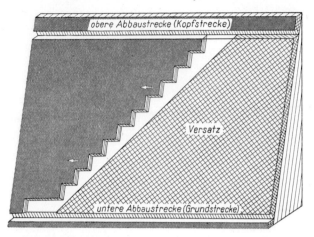

b) Cascade overhand stoping.

for: obere Abbaustrecke   read: upper mining level
     untere Abbaustrecke   lover mining level
     Versatz   fill (waste)

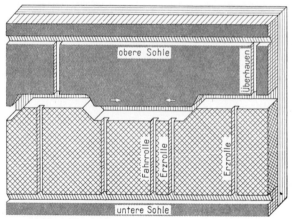

c) Overhand stoping, shrinkage stoping.

for: obere Sohle   read: upper level
     Überhauen   raise
     Erzrolle   ore chute
     Fahrròlle   manway
     untere Sohle   lower level

57

ness of the orebody. *Pillar caving:* Ore is broken in a series of stopes or tall rooms, leaving pillars between. Eventually the pillars are forced or allowed to cave under the weight of the roof.

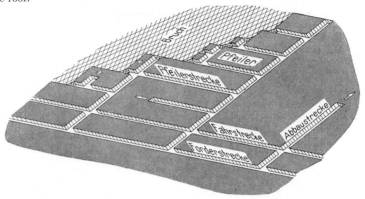

d) Room and pillar stoping.

for:	read:
Förderstrecke	haulage way
Fahrstrecke	manway
Abbaustrecke	drift
Pfeilerstrecke	crosscut
Pfeiler	pillar
Bruch	fill (waste)

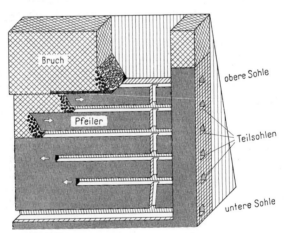

e) Sublevel stoping, cut and fill stoping.

for:	read:
obere Sohle	upper level
Teilsohlen	sublevels
Pfeiler	pillar (block)
untere Sohle	lower level
Bruch	fill (waste)

58

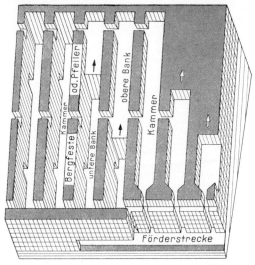

f) Room and pillar method.
   for: Kammer   read: room (chamber, stope)
        Bergfeste   pillar
        Förderstrecke   haulage level

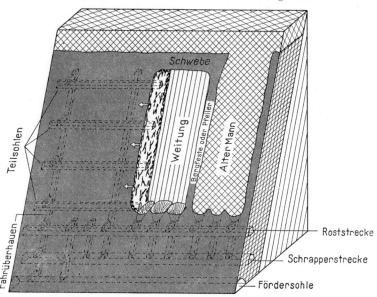

g) Open stope caving.
   for: Schwebe   read: roof
        Weitung   open stope

59

Bergfeste oder Pfeiler	pillar
Alter Mann	fill (waste)
Fahrüberhauen	manway raises
Teilsohlen	sublevels
Roststrecke	grizzly level
Schrapperstrecke	scrapping level
Hauptförderstrecke	main haulage level

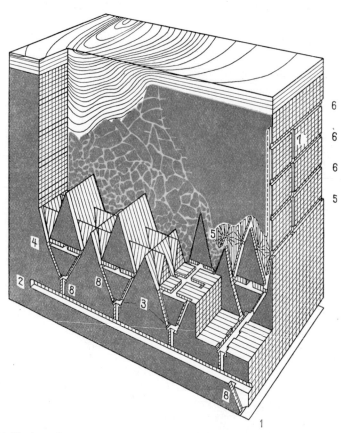

h) Block caving.

for:		read:	
1	Fördersohle		haulage level
2	Sammelschrapperstrecke		main scrapping level
3	Schrapperstrecken		scrapping levels
4	Roststrecken		grizzly levels
5	Unterschneidestrecken		undercuts
6	Schlitz- oder Kerbstrecken		crosscuts
7	Überhauen		raises
8	Erzrollen		ore chutes

60

Glory hole method: During mining, the ore is broken in a glory hole and dropped by gravity to a working level from where it is extracted and transported to the surface.

Hydraulic mining: Method of placer mining where layers of earth or gravel are broken up and washed into channels or sluices by sharp jets of water.

Longwall method: A system of working a seam of coal or layered ore in which the whole seam is taken out and no pillars are left (except shaft space pillars and sometimes the mainroad pillars).

Room and pillar method: A method of mining whereby the ore (or coal) is mined in a series of rooms leaving pillars between.

Shrinkage stoping: A modification of *overhand stoping* in which ore is mined beginning on a level and working upward, withdrawing only enough broken ore to leave a space between the broken ore and the back for the men at work. After the stope reaches the next level above (pillars being left to support the floor of the level), the broken ore is drawn from the stope.

Square-setting (Square-set stoping): Method of mining whereby the stope is filled with square-set timbering as the ore is removed.

Sublevel space stoping: Stoping by driving a series of sublevels, all advancing horizontally, in steplike arrangement and separated by horizontal pillars. Used in inclined deposits. As work advances the pillars are caved progressively, starting with the uppermost. The broken (caved) ore is removed and the waste is left.

Métodos de minería (los más importantes):

Hundimiento: Método de minería por el cual un bloque de mineral que sostiene a otro mucho mayor y que va a ser extraído, se deja caer o hundir; al caerse se rompe suficientemente para ser operable; el estrato superyacente se hunde a medida que el mineral se va retirando.

Variaciones: *Tajeo de arriba a abajo:* Se extrae una capa horizontal de mineral, permitiendo hundirse al bloque inmediato superior. Se deja entonces el bloque inmediato inferior temporalmente intacto hasta que el próximo bloque inferior se extraiga, permitiendo al bloque intermedio hundirse. El proceso se repite hasta alcanzar el fondo del depósito mineralizado. *Hundimiento en bloque:* Similar al anterior, excepto que el bloque que va a hundirse es mucho más grande y puede abarcar todo el depósito mineral. *Hundimiento en pilares:* El mineral se rompe en forma de tajos o cámaras, dejando pilares entre ellos. Eventualmente los pilares se hunden o se dejan caer bajo el peso del techo.

Método de embudo de extracción: Durante las operaciones mineras, el mineral es roto en un hoyo y dejado caer por gravedad a un nivel de trabajo de donde es extraído y transportado a la superficie.

Longwall: Sistema de trabajo en un manto de carbón o capa de mineral, en el cual todo el depósito se extrae sin dejar pilares (excepto pilares pilotos o pilares del camino principal).

Bovedones y pilares (room and pillar): Método de minería por medio del cual la mena (o carbón) se extrae mediante una serie de bovedones dejando pilares entre ellos.

Testeros con tolvas: Modificación de gradientes invertidos en los que la mena se extrae empezando en un nivel y trabajando hacia arriba, retirando sólo suficiente mena para dejar un espacio de trabajo entre la mena rota y el techo. Después que los trabajos han alcanzado el nivel inmediato superior (dejando pilares para soportar el piso del nivel), se extrae la mena rota del tajo.

Tajeo en cuadros: Método de minería por el cual la mena del tajo se reemplaza por cuadros de madera tan pronto como la mena se ha extraído.

Tajeo en subniveles: Explotación por medio de una serie de subniveles que avanzan horizontalmente, dispuestos como los peldaños de una escalera y separados por pilares horizontales. Usado en yacimientos inclinados. A medida que los trabajos avanzan se hunden progresivamente los pilares comenzando por los superiores. La mena rota se extrae y la roca se deja.

Méthodes minières (les plus importantes):

Foudroyage: Méthode d'exploitation dans laquelle le support d'un grand bloc ayant été supprimé, on laisse foudroyer ou tomber le minerai, qui en tombant se brise suffisamment pour pouvoir être manipulé. Il y a subsidence des couches surmontantes lorsqu'on soutire le minerai. Variations: *Tranche unidescendante foudroyée:* Le minerai est soutiré en tranches horizontales, prises en descendant. Chaque niveau est dépilé par petites sections, le toit de chaque section étant foudroyé avant que la section suivante soit attaquée. *Blocs foudroyés:* Méthode semblable à la tranche unidescendante foudroyée mais dont la tranche foudroyée est bien plus épaisse et peut même atteindre toute l'épaisseur du gîte. *Piliers foudroyés:* Le minerai est abattu dans une série de gradins ou de chambres hautes en laissant des piliers. Le cas échéant, les piliers foudroyent ou sont induits au foudroyage sous la pression du toit.
N. B. – En fait la définition de FAY correspond à la méthode dite des «chambres et piliers» *(Room and pillar stoping)* avec ou sans foudroyage ultérieur. Textuellement *«pillar caving»* désignerait le foudroyage des piliers, c'est-à-dire leur récupération.

–: Lors de l'abattage, le minerai tombe dans une grande cavité et suivant la gravité jusqu'à un niveau de roulage, d'où il est extrait et transporté à la surface.

Abattage hydraulique: Méthode d'exploitation de placers: les couches de sable et gravier sont désagrégées et entraînées dans des canaux ou des sluices par des jets d'eau très puissants (monitors).

Longue taille: Système d'exploitation d'une couche de charbon ou de minerai stratoïde dans laquelle la couche entière est défruitée et aucun pilier n'est laissé (excepté les stots de puits et quelquefois les stots de roulages principaux).

Chambres et piliers: Méthode d'exploitation par laquelle le minerai (ou le charbon) est abattu dans des chambres ménageant entre elles des piliers.

Chambre-magasin: Modification des gradins renversés dans laquelle le minerai est abattu en commençant à partir d'un niveau et en montant, en soutirant juste le minerai suffisant pour laisser un espace entre le minerai abattu et le front, permettant ainsi aux mineurs de travailler; lorsque la chambre atteint le niveau supérieur, des piliers étant laissés pour soutenir la sole de ce niveau, le minerai abattu est soutiré du magasin.

Chambre vide charpentée: Methode d'exploitation dans laquelle la chambre est remplie avec des cadres trirectangles au fur et à mesure que le minerai est enlevé.

Chambre magasin avec front vertical rabattant sur galérie horizontale préalable: Exploitation par chambre, avec une série de sous-niveaux, avançant horizontalement, disposés en gradins, et séparés par des piliers horizontaux. Utilisé dans des gîtes très inclinés. Au fur et à mesure de l'avancement des travaux, les piliers sont progressivement foudroyés en commençant par les plus hauts. Le minerai abattu (foudroyé) est soutiré et une partie est laissée en place.

Abbaumethoden (nur die wichtigsten):

Bruchbau: 1. Eine Abbaumethode, bei der das Erz durch Ausbrechen aus dem Verband gelöst wird, nachdem diesem die Unterlage genommen wurde, und bei der das Erz gleichzeitig in eine für die Handhabung günstige Korngröße zerbricht. 2. Eine Abbaumethode, bei der nach dem Entfernen des Erzes (Kohle) die überlagernden Schichten hereinbrechen und den entstandenen Hohlraum ausfüllen. Mit zunehmender Wiederverfestigung des gebrochenen Gesteins durch das Gewicht der überlagernden Schichten entstehen Senkungen an der Tagesoberfläche. Variationen: *Abwärts geführter Querbruchbau:* Eine horizontale Scheibe des Materials wird so hereingewonnen, daß das überlagernde Material nachbrechen kann. Die unmittelbar darunter liegende Scheibe wird jeweils übergangen und nur die nachfolgende Scheibe hereingewonnen, so daß die dazwischen liegende Scheibe nachbricht und abgefördert werden kann. Dieser Prozeß wird bis zum Grunde des Erzkörpers wiederholt. *Blockbruchbau:* Ähnlich wie oben, nur daß die hereinbrechende Scheibe viel höher ist und häufig den ganzen Erzkörper in seiner vertikalen Erstreckung ausmachen kann. Das Mineral wird so lange nach unten abgezogen, bis das den Erzkörper überlagernde Nebengestein zu den Ladestellen nachgerutscht ist. *Pfeilerbruchbau:* Das Mineral wird gleichzeitig in einer Reihe von benachbarten Abbauen verschiedenen Ausmaßes hereingewonnen, zwischen denen jeweils Bergfesten stehengelassen werden. Gelegentlich läßt man die Pfeiler bergtechnisch (Schießarbeit) oder unter dem anstehenden Gebirgsdruck zu Bruch gehen.

Trichterbau: Im Zuge des Abbaus wird das Erz gebrochen und es gelangt – der Schwerkraft folgend – zu einer im Tiefsten des Trichters liegenden Fördersohle; hier wird es geladen und abtransportiert.

Hydraulische Gewinnung: Methode des Abbaus von Seifen, bei der Erd- oder Kiesschichten durch einen druckstarken Wasserstrahl gelöst und in Kanäle oder Waschanlagen geschwemmt werden. Die Methode wird auch für die Kohlegewinnung unter Tage angewendet.

Streb: Bezeichnung für einen typischen Gewinnungsort in Kohlenflözen und geschichteten Erzlagerstätten. Der Streb ist ein langgestreckter Grubenbau von 150 bis 300 m Länge im Flöz oder Erz, an dessen Langfrontseite das Material zur Gewinnung ansteht und an dessen anderer Seite der entstandene Hohlraum verbricht oder mit minderwertigem, neu zugeführtem Material wieder verfüllt wird. Die gesamte Strebfront rückt täglich um einen Betrag von 1 bis 5 m zu Felde, so daß das Flöz oder Erz ohne Hinterlassung von Pfeilern gewonnen wird. (Ausnahmen gelten häufig für die Umgebung von Schächten und Hauptförderstrecken.)

Kammerpfeilerbau: Eine Gewinnungsmethode, bei der Erz, Salz oder Kohle in weiträumigen Bauen hereingewonnen wird, zwischen denen Pfeiler aus dem zu gewinnenden Mineral stehen bleiben.

Firstenstoßbau: Abbaumethoden in einer steilstehenden Ganglagerstätte, bei der der Gewinnungsvorgang von unten nach oben verläuft. Das hereingewonnene Material wird jedoch unten auf der Hauptsohle nur so weit abgezogen, daß die Bohr- und Schießarbeit vom Haufwerk aus durchgeführt werden kann. Wenn die Firste die nächst höhere Sohle oder den Pfeiler, der zum Schutz der höheren Sohle stehen geblieben ist, erreicht hat, wird das gesamte „magazinierte" Erz abgezogen. (Entwickelt sich aus dem Stoßbau, wenn das Einfallen steil ist.)

Blockbau mit Geviertzimmerung, Rahmenbauverfahren: Abbaumethode, bei der die Lagerstätte in einzelne Blöcke zerlegt und der jeweils ausgeerzte Raum mit Rücksicht auf das nur wenig standfeste Gebirge mit einem Geviertausbau aus Kanthölzern versehen wird.

Teilsohlenbruchbau, Teilsohlenbau: Pfeilerbau, bei welchem die Pfeiler übereinander aufgefahren werden und stehen. Bei fortschreitendem Abbau werden die Pfeiler nacheinander zurückgebaut, der oberste immer zuerst. Das Erz wird entfernt und das Gangmaterial und Gestein versetzt zur Stützung der aufgelassenen Wände („Alter Mann") wie beim Firstenstoßbau.

Mother lode:

See *lode*.

Criadero: Ver venero.

—: Voir lode.

Mutter- oder Hauptgang: Siehe Gang.

Muck

(Cornish; Mullock): 1. Useless material, such as earth, gravel, or barren rock. 2. To move waste or ore, usually with a shovel. 3. In some districts the term is used to denote ore. (McKinstry)

—: (Cornish mullock): 1. Material sin valor, como tierra, cascajo o roca estéril. 2. Mover lastre o mena con una pala. 3. En algunos distritos el término se usa para denotar mena.

1 – **Déblai:** Matériau sans valeur tel que terre, gravier ou roche stérile.

2 – **Mariner:** Charger stériles ou minerais, généralment à la pelle.

3 – **—:** Dans certains districts le minerai est désigné par le terme de *muck*.

Haufwerk: Nutzloses Material wie Erde, Kies, Schotter, taubes Gestein oder auch Erz, das frisch aus dem natürlichen Verband gewonnen wurde und seiner weiteren Bestimmung im Bergbaubetrieb zugeführt werden soll.

Neosom:

Added part in an invaded or metamorphosed rock or ore (opp. *paleosom*).

Neosoma: Parte agregada en una roca o mena invadida o metamorfizada (opuesto: paleosoma).

Néosome: Part étrangère d'une roche ou d'un minerai remplacé ou métamorphisé. (Contraire: paléosome.)

Neosom: Ein hinzugefügter Teil in einem metamorphen Gestein oder Erz. (Gegenteil: Paläosom).

Neptunism

(Neptunian theory): (Werner 1749–1807): The idea of Werner that practically all major rocks of the earth's crust were formed by sedimentation processes in oceans.

Neptunismo (Teoría Neptuniana): (Werner 1749–1807): al – Teoria de Werner según la cual todas las rocas mayores de la corteza de la tierra fueron formadas por procesos de sedimentación en los océanos.

Neptunisme (Werner 1749–1807): L'idée de Werner suivant laquelle presque toutes les roches importantes de l'écorce terrestre se sont formées par des processus sédimentaires dans les océans.

Neptunismus: Neptunistische Theorie (WERNER, 1749–1807): Die Vorstellung WERNERS, nach der praktisch alle weiter verbreiteten Gesteine der Erdkruste durch Sedimentationsprozesse in den Ozeanen gebildet wurden.

Nugget:
A large lump of placer gold. (SCH.)

Pepita: Pedazo grande de oro de placer.

Pépite: Gros morceau d'or de placer.

—: Ein ungewöhnlich großes Stück Seifengold.

Ophthalmites:
Consisting of coarsely lenticular (eye- *"augen"* shaped), boulder-like, or nodular elements in a ground mass.

Oftalmitas: Dícese de elementos de forma más o menos lenticular (forma de ojos *"augen"*), de canto rodado o nodular dentro de la matriz.

Roches oeillées: Il s'agit d'éléments grossièrement lenticulaires (en forme d'yeux ou «*augen*»), ou en forme de galets, ou de nodules dans une matrice.

Ophthalmit: Bezeichnung für grobe, linsenförmige („augenförmige"), geröllartige, oder knollige Stücke in einer Gesteinsmasse.

Ore:
Any rock mass from which one or more minerals or metals can be recovered at a profit. (Many materials from deposits of industrial minerals are also called ores.)

Bedded ore deposit: Ore aggregations in sedimentary rocks, exhibiting essentially the same bedded nature as the sedimentary country rock.

Cockade ore, Ring ore; Sphere ore: Ore formed by deposition of successive crusts of minerals around breccia fragments in a vein.

Drag ore: Fragments of ore torn from an ore body by faulting and scattered along the fault plane or zone.

Mena: Cualquier masa de roca de la cual uno a más minerales o metales pueden ser recuperados con utilidad. (Muchos materiales de depósitos de minerales industriales se llaman también menas.)

Depósitos minerales estratificados: Son agregaciones de menas en rocas sedimentarias, exhibiendo esencialmente la misma naturaleza estratificada que la roca sedimentaria encajonante. (Sinónimo: yacimientos minerales estratificados.)

Mena en cocardas: Mena anular; mena esferoidal – mena formada por la deposición sucesiva de costras de minerales alrededor de los fragmentos de brecha en una veta.

Mena de rastra: Fragmentos de mena rota de un depósito mineral por fallamiento y esparcidos a lo largo del plano o zona de falla.

Minerai: Toute roche à partir de laquelle on peut extraire avec bénéfice un ou plusieurs minéraux, un ou plusieurs métaux. (De nombreux matériaux d'origine minérale utilisés dans l'industrie, mais ne correspondant pas à la définition ci-dessus, sont cependant appelés minerais!)

Gîte de minerai stratiforme: Amas de minerai dans les roches sédimentaires particularisé par une disposition stratifiée identique à celle des roches sédimentaires encaissantes.

Minerai en cocarde: Minerai formé par le dépôt de croûtes minérales successives autour de fragments bréchiques dans un filon.

Minerai entraîné: Fragments de minerai arrachés à un corps minéralisé par un accident cassant et dispersés le long du plan de faille.

Erz: Jeder Mineralverband, der ein oder mehrere Minerale derart angereichert enthält, daß dieser zur technischen Gewinnung eines oder mehrerer Minerale oder Metalle verwendet werden kann.

Geschichtete Erzlagerstätte: Eine Erzanreicherung in Sedimentgestein, die grundsätzlich dieselbe Schichtung wie das Nebengestein aufweist.

Kokardenerz: Ringerz, Kugelerz − ein Erz, das durch die Anlagerung von aufeinanderfolgenden Krusten verschiedener Mineralarten um Brekzienzentren gebildet wurde.

Abgeschertes Erz, Ruschelerz: Bezeichnung für Bruchstücke von Erz, die bei Verwerfungsvorgängen von einem Erzkörper losgetrennt und entlang der Bruchfläche verschleppt wurden.

Ore bed:

Usually denoting an ore deposit of wide extent in sedimentary rock (synon. *ore seam*).

Mena estrato: Generalmente designa un depósito mineral de amplia extensión en roca sedimentaria (Sinónimo: yacimiento estrato).

Couche minéralisée: S'applique généralement à un corps minéralisé de grande extension dans les roches sédimentaires.

Flöz: Eine flächenhaft weitgestreckte, steil oder flach gelagerte Lagerstätte sedimentärer Entstehung mit Mächtigkeiten von einigen Zentimetern bis zu mehreren Metern (syn. Erzlager).

Ore deposit (see Figs. 8, 9):

Any mass of economic minerals or rocks, consolidated or non-consolidated, that is proved to be minable at a profit (see *ore*).

Depósito o yacimiento mineral: Cualquier masa de minerales económicos, consolidados o inconsolidados, susceptibles de ser explotados con utilidad (ver mena).

Gîte métallifère: Toute masse de minéraux utiles, consolidés ou non, susceptible d'être exploitée avec bénéfice (voir minerai).

Erzlagerstätte: Jede Anhäufung gewinnbarer Erzminerale (siehe Erz).

Ore dyke

(dike): An injected tabular mass of ore matter (presumably forced in a liquid or plastic state across the bedding or other layered structures of the invaded formation).

Mena dique: Intrusión tabulár (presumiblemente inyectado en estado líquido ó plástico a través de la estratificación u otras estructuras estratificadas en la formación invadida).

Dike de minerai: Intrusion de minerai intrusif de forme tabulaire. (Vraisemblable-

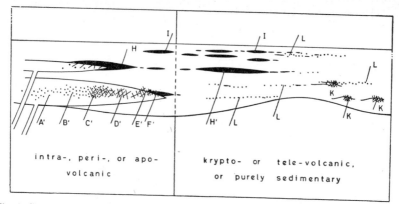

Fig. 8. Some basic ore patterns in and near extrusive metalliferous porphyries and in stratiform ore deposits.

(Normally with spilitic or propylitic "facies"); some typical examples: Rio Tinto belt, Spain-Portugal; volcanic belt from Ireland to Scotland to Norway; Appalachian and New Brunswick belt; Coast Range belt, Western U. S. A. Noranda belt, Quebec, and other Precambrian belts in shield areas, Ural Mts. belt.

A', B', C', D', E', F' correspond roughly to the intrusive patterns in porphyry coppers (see vein systems); and H and H' are massive deposits adjacent and somewhat removed from the extrusive rock; I are lenticules, layers or layered patches; K are late diagenetic compaction fractures in and near organic reefs; L represents disseminations. H', I, K, and L may also occur as euxenic sediments in no connection with extrusive rocks or exhalations. (Fig. 17 in: AMSTUTZ, G. C. and BUBENICEK, L. (1966) The diagenesis in sedimentary mineral deposits. In: Diagenesis, Elsevier, Amsterdam.)

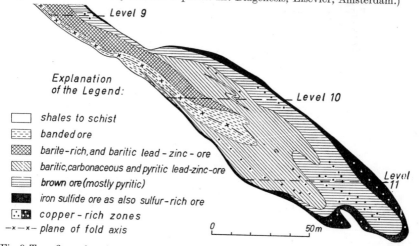

Explanation of the Legend:

- ☐ shales to schist
- ▦ banded ore
- ▨ barite-rich, and baritic lead – zinc – ore
- ▧ baritic, carbonaceous and pyritic lead-zinc-ore
- ▤ brown ore (mostly pyritic)
- ■ iron sulfide ore as also sulfur-rich ore
- ⬛ copper - rich zones
- –x–x– plane of fold axis

Fig. 9. Type figure for a folded stratiform deposit. (Profile of the "Neues Lager" of the Rammelsberg deposit in Germany; below Level 9, Coord. 1540/1550; After E. KRAUME, 1960).

67

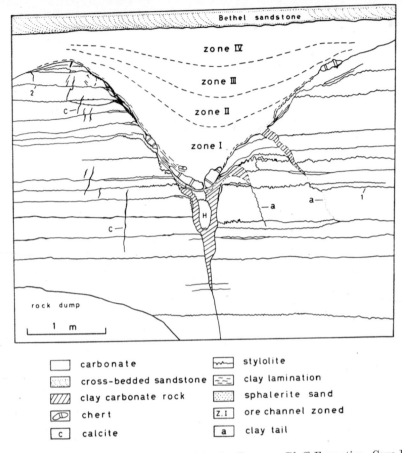

Fig. 10a. Sedimentary primary channel in the Downeys Bluff Formation, Cave-In-Rock fluorite district, southern Illinois. After PARK (1962).

ment injecté à l'état liquide ou plastique à travers le litage ou d'autres structures litées de la formation envahie).

Erzgang: Eine deszendente, aszendente oder lateralsekretionäre Erzeinlagerung mit flächenhafter unregelmäßiger Ausdehnung und erkennbarer Dicke (= Mächtigkeit). Es muß angenommen werden, daß die Massen in flüssigem, gasförmigem oder gelöstem Zustand die Schichtung des Gebirges durchkreuzt oder in die Gebirgsmasse eingedrungen sind.

Ore pipe:

Ore chimney: A vertically elongated body of ore, with a roughly circular or oval shaped cross section, often filled with breccia (breccia-pipe).

Mena tubular (Chimenea mineral): Masa de mineral alargada verticalmente, de forma aproximadamente circular u oval visto en corte transversal, frecuentemente rellenada con brecha (brecha tubular).

Cheminée minéralisée: Corps minéralisé allongé verticalement, de section horizontale grossièrement circulaire ou ovale, souvent empli par des brèches (cheminées bréchiques,)

Erzschlot: Ein steilstehender, schlotförmiger Erzkörper mit ungefähr kreisförmigem oder elliptischem Querschnitt; er ist häufig mit Brekzien gefüllt (Brekzien-Schlote).

Ore pocket (see Figs. 10a, b and 11):

Nest; bunch; kidney: Terms designating small irregular concentrations of ore. (Sch.)

Bolsas de mineral: Masas en forma de bolsa, racimo, riñón; son términos que designan pequeñas e irregulares concentraciones de mena.

Fig. 10b. Secondary solution cavities in limestone.

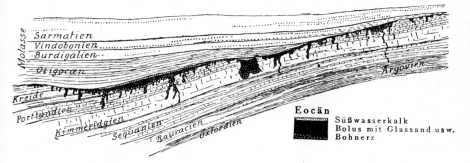

Fig. 11. Accumulation of ore matter in ore pockets (typical for certain Fe, Mn, and Bauxite deposits). Bean, pea, or pisolitic iron ore ("Bolus, Bohnerz") in the Jura Mountains of Switzerland and Southern Germany (after A. Heim, 1919).

Poche de minerai: Nid, mouche, reins: Termes désignant de petites concentrations irrégulières de minerai.

Erztasche: Nest, Bündel oder Niere: Bezeichnung für kleine unregelmäßig geformte Erzkörper.

Ore reserve classification (see Fig. 12):

Geometric (used in parts of Germany and Russia): see fig. 12. – American: 1. *Indicated ore* is ore for which tonnage and grade are computed partly from specific measurements, samples, or production data and partly from projection for a reasonable distance on geologic evidence. The sites available for inspection, measurement, and sampling are too widely or otherwise inappropriately spaced to outline the ore completely or to establish its grade throughout. 2. *Inferred ore* is ore for which quantitative estimates are based largely on broad knowledge of the geologic character of the deposit and for which there are few, if any, samples or measurements. The estimates are based on an assumed continuity or repetition for which there is geologic evidence; this evidence may include comparison with deposits of similar type. Bodies that are completely concealed may be included if there is specific geologic evidence of their presence. Estimates of inferred ore should include a statement of the special limits within which the inferred ore may lie. 3. *Measured ore* is ore for which tonnage is computed from dimensions revealed in outcrops, trenches, workings, and drill holes and for which the grade is computed from the results of detailed sampling. The sites for inspection, sampling, and measurement are so closely spaced and the geological character is so well defined that the size, shape, and mineral content are well established. The computed tonnage and grade are judged to be accurate within limits which are stated, and no such limit is judged to differ from the computed tonnage or grade by more than 20 per cent. (U. S. Bureau of Mines and U. S. Geol. Surv., cf. PARKS)

Clasificación de reservas de mineral: Geométrica (en partes de Alemania y Rusia); ver Fig. 12 – Americana: 1. *Mineral indicado* es el mineral en el cual el tonelaje y la ley se calculan en parte por mediciones específicas, muestras o datos de producción, y en parte por la proyección de una distancia razonable basada en evidencia geológica. Los lugares accesibles para inspección, medición o muestreo están demasiado distantes o de lo contrario inapropiadamente separados para delinear con exactitud el mineral o para establecer enteramente su ley. 2. *Mineral inferido* es aquel en el cual una estimación cuantitativa se basa considerablemente en un amplio conocimiento del carácter geológico del depósito y para el cual sólo se dispone, si se dispone de alguna, de unas pocas muestras y medidas. La estimación está basada en una supuesta continuidad o repetición para lo cual existe una evidencia geológica; esta evidencia puede comprender comparación con otros depósitos de tipo similar. Cuerpos de mineral completamente ocultos pueden ser incluidos si existe evidencia geológica específica de su presencia. La estimación del mineral inferido podría comprender una exposición de sus límites especiales dentro del cual el mineral inferido podría estar ubicado. 3. *Mineral medido* es aquel cuyo tonelaje se calcula de acuerdo con las dimensiones reveladas en los afloramientos, trincheras, labores y perforaciones diamantinas, y en el cual la ley se calcula por los resultados de un muestreo detallado. Los lugares de inspección, muestreo y mediciones están cercanos y el carácter geológico suficientemente bien definido de modo que el tamaño, forma y contenido del mineral sean bien establecidos. Se considera que los cálculos de ley y tonelaje son exactos dentro de los límites establecidos y no pueden tales límites diferir del tonelaje y ley calculados en más de un 20%.

Classification des réserves minières: Géométrique (en R.D.A. et en U.R.S.S.); voir Fig. 12.
– Américaine: 1 – *Minerai possible*. Le tonnage et la teneur sont estimés en partie par analyses, échantillonnage ou résultats de production, en partie par extrapolation sur une distance raisonnable de données géologiques. Les endroits accessibles en vue d'observation, de mesure, de prélèvements d'échantillons s'avèrent trop ou mal espacés pour circonscrire complètement le corps minéralisé ou estimer la distribution de ses teneurs. 2 – *Minerai probable*. L'estimation quantitative repose grandement sur les connaissances d'ensemble qu'on a des caractères géologiques du gisement pour lequel on ne dispose que de peu, s'il y en a, d'échantillons ou de mesures. L'estimation repose sur une continuité ou une répétition supposées du corps minéralisé, ces suppositions étant justifiées par des connaissances géologiques parmi lesquelles la comparaison avec des types analogues de gisement n'est pas exclue. Des corps minéralisés complètement enfouis peuvent être compris dans cette évaluation s'il existe des présomptions géologiques suffisantes. L'estimation des réserves probables devrait comporter la définition des limites à l'intérieur desquelles le minerai probable doit se trouver. 3 – *Minerai reconnu*. Le tonnage est estimé à partir des affleurements, tranchées, travaux miniers, forages sur lesquels des échantillonnages systématiques ont été effectués. Les endroits de prélèvements, de mesure et de reconnaissance s'avèrent suffisamment bien définis pour que les dimensions, les formes et le contenu du corps minéralisé soient bien établis. Le tonnage et les teneurs à l'intérieur des limites définies sont supposés avoir été estimés de manière précise, c'est-à-dire que les limites reconnues ne doivent pas s'éloigner de plus de 20 pour cent des tonnage et teneur estimés.

Française: Classification française: il n'y a pas, à proprement parler, de conception purement française en ce domaine, la classification américaine étant couramment employée. Il convient cependant de citer ici la proposition de BLONDEL et LASKY (1956):

Ressources = Réserves + minerai potentiel

Le dernier terme peut lui-même se décomposer en fonction des conditions économiques et techniques, de sorte que l'on peut écrire:

Ressources = Réserves + ressources marginales + ressources submarginales + ressources latentes.

Ces auteurs retiennent les catégories de réserves prouvées, probables et possibles, mais selon les définitions de LEITH, plus souples que celles généralement adoptées. En échange ils proposent de modifier la classification des services publics U. S. en gr adant deux catégories seulement: réserves démontrées (groupant les réserves mesurées et les réserves indiquées) et inférées.

Klassifikation der Erzvorräte: *Vorratsklassen:* Geometrische Klassifikation (z. T. in Deutschland und der USSR gebraucht), siehe Fig. 12. – Amerikanisch: 1. *Indicated Ore:* Die Bezeichnung für Erze, deren Menge und Gehalte durch spezielle Messungen berechnet werden – aus Proben- oder Produktionsdaten und teilweise auch bis zu einer zuverlässigen Entfernung aus Projektionen, die auf geologischen Vorkommen aufbauen. Die Punkte der Einsichtnahme, Messungen und Probenahme sind zu verstreut oder zu unregelmäßig verteilt, um den Gehalt mit Sicherheit durch die ganze betrachtete Zone feststellen zu können. 2. *Inferred ore:* Die Bezeichnung für Erze, für die die quantitativen Bestimmungen größtenteils auf der Kenntnis des geologischen Charakters der Lagerstätte basieren und für die nur wenige – wenn überhaupt – Proben oder Messungen möglich sind. Die Schätzungen stützen sich auf eine angenommene Gleichmäßigkeit und Wiederholung, für die bestimmte geologische Anhaltspunkte bestehen. Diese Anhaltspunkte können auch Vergleiche mit ähnlich gearteten Lagerstätten einschließen. Erzkörper, die völlig unerschlossen sind, können dann eingereiht werden, wenn spezielle geologische Hinweise für ihre Existenz bestehen. Bestimmungen von „inferred ore" sollen eine Feststellung der Grenzen, innerhalb derer das Vorkommen liegen kann, beinhalten. 3. *Measured Ore*: Die Bezeichnung für Erzvorkommen, deren Mineralgehalt durch eingehende Probenahme festgestellt und deren Vorräte in Tonnen über die Ausdehnung berechnet wurden, die wiederum durch Ausbisse, Grubenbaue und Bohrlöcher zu ermitteln sind. Die Punkte für die Untersuchung, die Probenahme und die Messungen sollen so dicht beieinander liegen und der geologische Aufbau so gut bekannt sein, daß Größe, Form und Mineralzusammensetzung des Vorkommens hinreichend genau bestimmt werden können. Die errechneten Ausmaße und Gehalte werden innerhalb vorher festgelegter Genauigkeitsgrenzen angegeben, wobei die Grenze 20% nicht überschreiten soll.

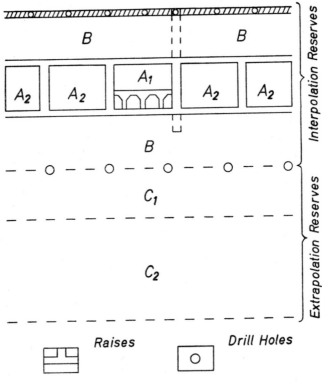

Fig. 12. Russian Classification of Ore Reserves (in plane of a vein) (After V. I. Smirnov, 1961, and F. Stammberger, 1956).

Category A_1: reserves fully studied and outlined by preparatory shafts or borings of operational prospecting along not less than 4 sides; hydrogeological conditions of mining have been studied; economic grades of useful minerals and their distribution have been established in each block; the nature and technology of treating useful minerals have been studied on the basis of experience in industrial utilization.

Category A_2: reserves have been prospected in detail and outlined by shafts or borings along not less than 3 sides; conditions of deposition, interrelationships of natural types, and economic grades of useful mineral, as well as the hydrological conditions of the deposit and the conditions of its development, have been studied, the grade and technological nature of the useful mineral have been determined to a degree of detail which makes possible the planning of the treatment and technology for utilizing the useful mineral.

Category B: reserves have been prospected and outlined by shafts or borings along not less than 2 sides; conditions of deposition have been studied, natural types and economic grades of useful mineral have been established without detailing their distribution; the grade and technological properties of the useful mineral have been studied to a degree which makes possible a choice of scheme for its treatment; general conditions of development, as well as the general hydrological conditions of the deposit, have been rather fully determined.

Category C_1: reserves have been determined on the basis of a sparse pattern of borings or shafts; reserves adjoining the limits of reserves of categories A_1, A_2, and B; reserves of particularly complex deposits, for which, in spite of a close pattern of exploratory shafts, the distribution of the valuable component or mineral has not been determined; the quality, natural type, economic grade and technological treatment of the useful mineral have been tentatively determined on the basis of analyses or laboratory tests of samples taken, and also by analogy with deposits already studied; general conditions of development, as well as the general hydrological conditions of the deposit, have been tentatively studied.

Category C_2: reserves of adjoining segments of deposits prospected according to categories A_2, B and C_1, and also reserves assumed to exist on the basis of geological and geophysical data confirmed by testing the useful mineral in particular borings and shafts.

Fig. 12. Clasificación (alemán y ruso) de reservas de mineral (en el plano de la veta) (segun V. I. Smirnov, 1961, y F. Stammberger, 1956).

Categoría A_1: reservas estudiadas totalmente y delimitadas por trabajos mineros preparatorios o perforaciones para prospección (delimitados por 4 lados); las condiciones hidrogeológicas mineras son conocidas; la ley del mineral útil y su distribución son establecidas en cada bloque; la tecnología para el tratamiento del mineral útil es estudiada sobre la base de experiencias en utilización industrial.

Categoría A_2: las reservas son prospectadas en detalle y delimitadas por trabajos mineros o perforaciones (delimitados por lo menos por 3 lados); condiciones de depositación, distribución de la materia prima y ley del mineral útil, así como las condiciones hidrogeológicas del yacimiento y su desarrollo, son estudiadas. El valor y naturaleza tecnológica del mineral útil son determinados con el grado de detalle que hace posible planear el tratamiento y tecnología, para la utilización del mismo.

Categoría B: las reservas son prospectadas y delimitadas por lo menos por 2 lados, por trabajos mineros o perforaciones; condiciones de depositación son estudiadas; materia prima, ley del mineral útil son establecidas sin detalle de la distribución; el valor y propiedades tecnológicas del mineral útil son estudiados con un grado que hace posible la elección de un esquema para su tratamiento; condiciones generales de desarrollo así como las condiciones generales hidrogeológicas del yacimiento, son casi totalmente determinadas.

Categoría C_1: las reservas son determinadas sobre la base de una red mas espaciada de trabajos mineros o perforaciones; a) las reservas adyacentes a los límites de las categorías A_1, A_2 y B; b) reservas de yacimientos particularmente complejos, por los cuales, a pesar de una red densa de trabajos de exploración, la distribución del mineral útil no ha sido determinado. La calidad, naturaleza, ley y tratamiento tecnológico del mineral útil han sido provisionalmente determinados sobre la base de análisis o pruebas de laboratorio, de muestras tomadas y también por analogía con yacimientos ya estudiados. Condiciones generales de desarrollo, así como las condiciones hidrogeológicas del yacimiento son provisionalmente estudiadas.

Categoría C_2: reservas de segmentos de yacimientos prospectados adyacentes a las categorías A_2, B y C_1, y también reservas que se supone que existen, sobre la base de datos geológicos y geofísicos, confirmados por pruebas del mineral útil, en labores y perforaciones especiales.

Fig. 12 Classification des réserves minières: Les réserves des gites minéraux appartiennent à la:

– *classe A_1* si elles ont été complètement reconnues. Cela suppose que:

– la délimination soit faite sur au moins de quatre côtés, par des travaux miniers ou des sondages faits lors de la reconnaissance détaillée au cours de l'exploitation,

– les qualités de la matière première, ses différentes variétés et leur répartition à l'intérieur de chaque unité d'exploitation aient été reconnues à un degré tel que pratiquement on peut s'attendre à ce qu'il n'y ait pas de différences entre les réserves calculées à l'avance et les résultats effectivement obtenus lors de l'exploitation,

– les conditions hydrogéologiques et les conditions géologiques pour l'exploitation aient été reconnues,

– la technologie du traitement de la matière première soit connue par expérience industrielle.

– *classe A_2* si elles ont été reconnues en détail. Cela suppose que:

– elles soient délimitées par des travaux miniers ou des sondages sur au moins de trois côtés,

– les conditions du gisement et la tectonique, les qualités de la matière première, la répartition des différentes variétés de celle-ci, les conditions géologiques et hydrogéologiques du gisement en vue de la préparation des réserves aient été étudiées,

– les propriétés technologiques de la matière première soient connues au point qu'un projet technique pour la suite des opérations puisse être élaboré.

– *classe B*, si elles ont été suffisamment reconnues. Cela suppose que:

– elles soient délimitées par des travaux miniers ou des sondages sur au moins de deux côtés,

– les conditions de gisement et les grands traits de la tectonique aient été étudiés, les qualités et variétés de la matière première aient été constatées, à l'exception de détails concernant leur répartition à l'intérieur du gisement,

– les conditions hydrogéologiques, de même que les conditions géologiques d'exploitation du gisement soient éclaircies dans leurs grands traits,

– les propriétés technologiques de la matière première aient été étudiées à un point tel qu'elles autorisent le choix d'un schéma pour le traitement ultérieur.

– *classe C_1*, si elles ont été reconnues dans leurs grands traits. Cela suppose que:

– elles aient été reconnues par un réseau à maille large de sondages ou de travaux miniers,

– la qualité de la matière première et ses variétés, les conditions géologiques générales et hydrogéologiques du gisement aient été étudiées dans leurs grandes lignes.

– la technologie du traitement ait été arrêtée, soit sur la base d'analyses, respectivement de vérifications de laboratoire, d'échantillons prélevés selon les règles ou par analogie avec des gisements déjà étudiés. (Dans le cas de réserves «hors bilan», dont la teneur est inférieure à la teneur limite économique, la technologie du traitement ultérieur peut encore rester inconnue).

D'autre part, il faut inclure ici des réserves qui

$1°$ – sont contiguës de réserves appartenant aux classes A_1, A_2 ou B

$2°$ – font partie de gîtes particulièrement compliqués et pour lesquels la répartition des composants économiquement intéressants ou du minerai n'a pas été éclaircie en dépit d'un réseau de travaux de reconnaissance à maille sérrée.

– *classe* C_2, quand elles ont été soupçonnées d'après les indications géologiques ou géophysiques et qu'elles ont été vérifiées par un échantillonnage de la matière première minérale dans des sondages isolés, des tranchées ou des affleurements ou quand elles sont mitoyennes de parties de gisements contenant des réserves des classes A_2, B ou C_1. Le type minéralogique et la qualité de la matière première doivent être déterminés dans des districts nouveaux par l'étude d'échantillons isolés ou de prélèvements. Lorsqu'il s'agit de l'extension d'un gisement connu, elles peuvent, au même titre que les conditions géologiques générales ou les conditions hydrogéologiques, avoir été admises sur une base d'analogies, respectivement avoir été montrées théoriquement et en ce qui concerne les réserves hors bilan encore être inconnues.

Fig. 12. Vorratsklassen: Geometrische Klassifikation (z. T. in Deutschland und evtl. der UdSSR gebraucht). (Nach V. I. SMIRNOV, 1961, und F. STAMMBERGER, 1956).

Klasse A_1, wenn sie vollständig erkundet, d. h. von bergmännischen Arbeiten oder durch Bohrungen der betrieblichen Nacherkundung an nicht weniger als vier Seiten begrenzt sind, die Qualität des Mineralrohstoffes, seine verschiedenen Arten und ihre Verteilung in jedem Abbaublock so weit festgestellt sind, daß Abweichungen zwischen den im voraus berechneten Vorratsmengen und den tatsächlichen Ergebnissen beim Abbau praktisch nicht zu erwarten sind, die hydrogeologischen und anderen geologischen Bedingungen für den Abbau erforscht sind und die Technologie der Verarbeitung des mineralischen Rohstoffes auf Grund industrieller Erfahrungen bekannt ist;

Klasse A_2, wenn sie eingehend erkundet, d. h. durch bergmännische Arbeiten oder Bohrungen an nicht weniger als drei Seiten begrenzt sind, die Lagerungsverhältnisse und Tektonik, die Qualitäten des Rohstoffes, die Verteilung der verschiedenen Rohstoffarten, die geologischen und hydrogeologischen Verhältnisse der Lagerstätte für die Aus- und Vorrichtung der Vorräte so erforscht und die technologischen Eigenschaften des Mineralrohstoffes so geklärt sind, daß das technische Projekt für die Weiterverarbeitung ausgearbeitet werden kann;

Klasse B, wenn sie hinreichend erkundet, d. h. durch bergmännische Arbeiten oder Bohrungen an nicht weniger als zwei Seiten begrenzt sind, die Lagerungsverhältnisse und die Grundzüge der Tektonik erforscht, die Qualitäten und Rohstoffarten bis auf Einzelheiten ihrer Verteilung für die Lagerstätte festgestellt sind, die hydrogeologischen Verhältnisse sowie geologischen Abbaubedingungen der Lagerstätte im allgemeinen geklärt, die technologischen Eigenschaften des Mineralrohstoffes so weit erforscht sind, daß die Wahl eines Schemas der Weiterverarbeitung gewährleistet ist;

Klasse C_1, wenn sie weiträumig erkundet, d. h. durch ein weitmaschiges Netz von Bohrungen oder bergmännischen Bauen begrenzt sind, die Qualität des Rohstoffes und seine Arten, die allgemeinen geologischen und hydrogeologischen Verhältnisse der Lagerstätte in ihren Grundzügen erforscht, die Technologie der Weiterverarbeitung entweder auf Grund von Analysen bzw. Laboratoriumsprüfungen ordnungsgemäß entnommener Proben oder in Analogie zu erforschten Lagerstätten bestimmt ist. (Bei Außerbilanzvorräten kann die Technologie der Weiterverarbeitung noch unbekannt sein).

Außerdem gehören hierzu Vorräte,
1. die an Vorräte der Klassen A_1, A_2 oder B angrenzen oder
2. Vorräte besonders komplizierter Lagerstätten, für die trotz eines engmaschigen Netzes von Erkundungsarbeiten die Verteilung der wertvollen Komponenten oder des Minerals nicht geklärt ist.

Klasse C_2, wenn sie nach geologischen oder geophysikalischen Angaben vermutet und durch Bemusterung des Mineralrohstoffes in einzelnen Bohrungen, Schürfen oder Ausbissen bestätigt wurden oder an Lagerstättenteile mit Vorräten der Klassen A_2, B oder C_1 angrenzen. Mineralart und Güte des Rohstoffes müssen in neuen Feldern durch Untersuchung einzelner Handstücke oder Proben bestimmt sein. Im Anschluß an bekannte Lagerstätten können sie ebenso wie die allgemeinen geologischen und hydrogeologischen Verhältnisse der Lagerstätten auf Grund von Analogien angenommen bzw. theoretisch aufgezeigt bzw. für Außerbilanzvorräte noch unbekannt sein."

Auszug aus F. STAMMBERGER (1956) Einführung in die Berechnung von Lagerstättenvorräten fester mineralischer Rohstoffe. Akademie-Verlag, Berlin, 153 Seiten (S. 6–8).

Ore shoot:

Part of a deposit in which the valuable minerals are more richly concentrated. Or; a high-grade concentration of ore along a fissure vein. (SCH.)

Clavo, bolsonada: Parte de un depósito en el cual los minerales económicos están más concentrados. O tambien una alta concentración de mineral a lo largo de una veta de fisura.

Colonne minéralisée: Partie d'un gisement dans laquelle les minéraux utiles sont plus richement concentrés; ou concentration à forte teneur de minerai dans le plan d'un filon.

Adelszone: Der Teil einer Lagerstätte, in dem das wertvolle Mineral überdurchschnittlich angereichert ist.

Ore sill (compare sill):

Intrusive ore sheet: A tabular sheet of magmatic ore (presumably injected in a liquid state along the bedding planes of a sedimentary or layered igneous formation). (SCH.)

Ore sill: Mena intrusiva laminar. Mena magmática de forma tabular (presumiblemente inyectada en estado líquido a lo largo de los planos de estratificación sedimentaria de una formación ígnea estratificada).

Sill de minerai: Corps minéralisé plat, intrusif. Minerai magmatique de forme plate et tabulaire (vraisemblablement injecté à l'état liquide entre les plans de stratification de formations sédimentaires ou ignées stratifiées).

Lagergang: Intrusives Erzlager von flächenhafter Ausdehnung, das sich aus magmatischen Substanzen gebildet hat und parallel zur Schichtung des Nebengebirges eingelagert ist. (Es wird angenommen, daß die Masse in flüssigem Zustand in die Trennflächen von Sedimenten oder geschichteten Erstarrungsgesteinen eingedrungen ist.)

Ore vein:

Metalliferous vein; (lode s. l.): A tabular or sheet-like mass of ore minerals occupying a fissure or a set of fissures and later, in formation, than the enclosing rock (see *vein*).

Veta mineral: Una veta metalífera; (filòn s. l.) – una masa tabular o laminar de minerales que ocupan una fisura o serie de fisuras y que son posteriores a la roca de caja (ver veta).

Filon minéralisé: Filón metallifère; (*lode* s. l.) – masse tabulaire ou plate de minéraux utiles occupant une fracture ou un groupe de fractures et de formation postérieure aux roches encaissantes (voir filon).

Erzgang: Erzhaltiger Gang (siehe Gang).

Paleosom:

Essentially unchanged original parts of an invaded or metamorphosed rock (opp. *neosom*).

Paleosoma: Esencialmente las partes originarias inalterables de una roca invadida o metamorfizada (opuesto *neosoma*).

Paléosome, ou roche hôte: Partie originelle essentiellement invariante dans une roche métamorphisée ou ayant subi un apport. (Contraire: *néosome*).

Paläosom: Die im wesentlichen unveränderten Teile eines eingedrungenen oder metamorphen Gesteins. (Gegenteil: *Neosom*).

Palingenesis (App. VIId):

The formation of a magma *in situ* by an ultrametamorphic process of partial anatexis or complete (regional) fusion or solution of deep-seated rocks.

Palingenesis: La fusión parcial o completa de una roca o la formación de un magma *in situ* debido a un proceso ultrametamórfico de anatexis parcial o completa (regional) fusión o solución de rocas profundas.

Palingenèse: Formation *in situ* d'une roche fondue ou même d'un magma par des processus ultra-métamorphiques d'anatexie partielle ou de fusion et mise en solution complètes (régionales) de roches, en profondeur.

Palingenese: Die Bildung einer Schmelze oder eines Magmas *in situ* durch ultrametamorphe Vorgänge, d. h. durch Anatexis oder vollständige Aufschmelzung von Gestein in der Tiefe.

Paragenesis (see Fig. 13 and Appendix VIII):

1. The assemblage of minerals which occur together (common meaning in European literature). 2. The order of deposition or crystallization of the minerals present (common meaning in American literature).

Paragenesis: 1. Un grupo de minerales que se encuentran juntos (significado común en la bibliografía europea). 2. El orden de deposición o cristalización de los minerales presentes (significado común en la bibliografía americana).

Paragenèse: 1 – Association de minéraux qui se rencontrent ensemble dans la nature (acception courante en littérature européenne). 2 – Ordre de dépôt ou de cristallisation des minéraux présents dans une association naturelle (acception courante en littérature américaine).

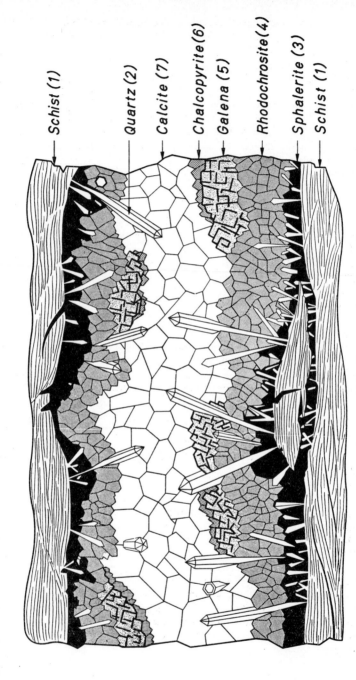

Schist (1)
Quartz (2)
Calcite (7)
Chalcopyrite(6)
Galena (5)
Rhodochrosite(4)
Sphalerite (3)
Schist (1)

Fig. 13. Paragenetic sequence (1–7) in a hydrothermal vein (after W. MAUCHER, 1914).

Paragenese: 1. Eine Familie von Mineralien, die zusammen vorkommen (übliche Bedeutung in der europäischen Literatur). 2. Die Reihenfolge der Ablagerung oder Kristallisation von betrachteten Mineralien (übliche Bedeutung in der amerikanischen Literatur).

Perimagmatic deposit (see Appendix VIIb):
formed at or near the wall or contact of a magmatic intrusion.

Depósito perimagmático: Formado en contacto con una intrusión magmática o cerca de sus paredes.

Gisement périmagmatique: Gisement formé au contact ou près du contact d'une intrusion magmatique.

Perimagmatische Lagerstätte: Eine Lagerstätte, die in der Nähe oder unmittelbar am Kontakt einer magmatischen Intrusion gebildet wurde.

Pillar caving:
See *caving*.

Hundimiento en pilares: Ver hundimiento.

Piliers foudroyés: Voir foudroyage.

Pfeilerbruchbau: Siehe Bruchbau.

Pinch:
A local thin place in an ore vein, an ore bed, or a mineral zone. (Sch.)

Estrechamiento: Adelgazamiento local de una veta mineral, de una mena estratificada o de una zona de mineral.

Pincement, étranglement: Zone d'amincissement dans un filon, une couche ou un amas minéralisé.

–, Eine örtlich begrenzte Verdünnung eines Ganges, einer geschichteten Lagerstätte oder einer mineralisierten Zone.

Pitch
(Pitches): 1. Ore pitch: The crosscutting connections of flats (pitches and flats are terms applied often in the Northern Tri-State area, Central and S. America). 2. Structural term: The angle between the axis of the ore shoot and the strike of the vein. The pitch is measured in the plane of the vein. (McKinstry)

Pitch: 1. Conexiones transversales de los *flats* (*pitches* y *flats* son términos frecuentemente usados en el área Tri-State del Norte, en América Central y del Sur. 2. El ángulo entre el eje de la bolsonada y el rumbo de la veta. El *pitch* está medido en el plano de la veta.

–: 1 – Bretelle: Connections sécantes entre lentilles minéralisées concordantes avec la stratification des roches encaissantes (les *pitches* et *flats* sont des termes fréquemment utilisés dans la partie septentrionale du district du Tri-State, en Amérique Centrale et du Sud. 2 – Chute: L'angle entre l'axe de la colonne minéralisée et de la direction du filon. Le *pitch* est mesuré sur le plan du filon.

79

—: 1. Die Querverbindungen zwischen mineralisierten Horizonten *(flats)*; die Ausdrücke sind vor allem im Northern Tri-State Gebiet, Zentral- und Südamerika gebräuchlich. 2. Der Winkel zwischen der Adelszone und dem Streichen des Erzganges. Der *Pitch* wird in der Ebene des Erzganges gemessen.

Placers (see Fig. 14):

Deposits of minerals (heavy or light) formed by moving water, air, or talus (cf. *eluvial, alluvial, diluvial, colluvial, eolian placers*).

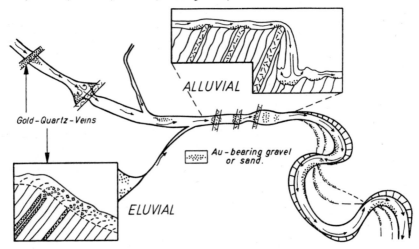

Fig. 14. Type pattern of alluvial and eluvial placer deposits.

Placeres: Depósitos minerales (pesados o ligeros) formados por movimientos de agua, aire o talud (ver placeres eluviales, aluviales, diluviales, coluviales, eólicos).

Placer: Gisement de minéraux (lourds ou légers) formés par des mouvements d'eau, d'air ou de talus. (Voir placer alluvial, colluvial, diluvial, éluvial, éolien).

Seifen: Mineralvorkommen, die durch Bewegungen des Wassers, der Luft oder des Gehängeschuttes gebildet wurden (z. B. eluviale, alluviale, diluviale, colluviale, litorale oder äolische Seifen).

Plan:

Representation of features such as mine workings or geological structures on a horizontal plane. (McKinstry)

Plano: Representación de conjuntos geométricos, tales como trabajos mineros o estructuras geológicas sobre un plano horizontal.

Plan: Représentation d'ensembles géométriques tels que travaux miniers ou structures géologiques sur un plan horizontal.

Riß: Im Vermessungswesen; Wiedergabe von Konturen — z. B. Grubenbauen oder geologischen Strukturen — in einer Horizontalebene.

Playa:

Playa: Closed basin in arid area, often with salt lakes or salt flats, which are the *playa deposits.*

Playa: Hondonada cerrada en un área árida frecuentemente con lagunas o bancos de sal que constituyen depósitos de playa.

Chott: Bassin fermé dans une région aride, comportant souvent des lacs salés ou des terrasses salées, qui sont les gisements de chott.

Schott: Abgeschlossenes Becken in ariden Gebieten, in denen Salzseen und Salzablagerungen vorkommen.

Plugging:

Sealing up a hole that was dry and is to be abandoned. It is generally done under governmental regulations, so that deep saline waters do not enter surface-water wells or contaminate other reservoirs in the vicinity. (LEVORSEN)

Taponamiento: La obturación de un hueco que ha estado seco y que va a ser abandonado. Se hace generalmente bajo regulaciones gubernamentales de modo que las aguas salobres profundas no se mezclen con las aguas dulces o se contaminen con otros reservorios de la vecindad.

Obturation: Fermeture d'un puits (ou d'un forage) sec qui doit être abandonné. L'obturation est réglementée, ceci afin que les eaux salines profondes ne contaminent pas les puits de surface ainsi que les roches réservoirs du voisinage.

—: Abdichtung eines nicht fündigen Bohrlochs, das aufgegeben werden soll. Die Abdichtung erfolgt meist unter staatlicher Aufsicht und nach Vorschriften, so daß zusitzende Salzwässer nicht in oberflächennahe Brunnen gelangen oder andere benachbarte Reservoire verunreinigen können.

Plunge:

The angle between any inclined line and a horizontal plane. The angle is always measured in a vertical plane containing the line. The term is used to designate the inclination of the axis of an oreshoot, the axial line of a fold, the attitude of lineation, etc. (Not to be confused with *Pitch* or *Dip*) (McKINSTRY).

Inclinación lineál: El ángulo entre cualquier línea y un plano horizontal. El ángulo siempre se mide en un plano vertical que contiene a la línea. El término se usa para designar la inclinación del eje de un clavo mineral *(oreshoot)*, la línea axial de un plegamiento o posiciones de lineamiento, etc.

Plongement: Angle entre une ligne inclinée et un plan horizontal. Cet angle est toujours mesuré dans le plan vertical contenant la ligne. Le terme est utilisé pour repérer l'axe d'une colonne minéralisée, l'axe d'un pli, la disposition d'une linéation, etc.

Lineares Einfallen: Der Winkel zwischen jeder geneigten Linie und der Horizontalebene. Der Winkel wird immer in einer Vertikalebene gemessen, der die geneigte Linie enthält. Der Ausdruck wird verwendet, um die Neigung von Erzgängen, von Faltenachsen und linearen Gefügen zu bestimmen (nicht zu verwechseln mit dem englischen Dip, Pitch oder Rake).

Plutonic (see Appendix VII b):

Of igneous intrusive origin. Usually applied to sizable bodies of intrusive rock rather than to small dikes or sills. (McKINSTRY)

Plutónico: De origen intrusivo ígneo. Generalmente aplicado a grandes cuerpos de roca intrusiva, más que a pequeños diques verticales u horizontales.

Plutonique: D'origine ignée et intrusive. S'applique généralement aux corps de roche intrusive d'une certaine dimension plutôt qu'aux dikes et aux sills.

Plutonisch: Gesteine intrusiven Ursprungs; gewöhnlich werden große intrusive Gesteinskörper eher plutonisch genannt als kleinere Kluft- und Gangfüllungen.

Plutonism

(Vulcanian, Plutonic theory): The idea of HUTTON (1726–1797) that magma is of great influence on the events in the earth's crust, in distinction to neptunism. (SCH.)

Plutonismo (Teoría plutónica, volcánica): Idea de HUTTON (1726–1797) según la cual el magma tiene gran influencia en los acontecimientos de la corteza terrestre, a diferencia del neptunismo.

Plutonisme (Théorie plutonique, volcanique): Idée de HUTTON (1726–1797) suivant laquelle le magma a un rôle primordial dans l'histoire de la croûte terrestre; elle s'oppose au neptunisme.

Plutonismus: Vulkanische oder Plutonische Theorie; Der Gedanke von HUTTON (1726 bis 1797), daß das Magma großen Einfluß auf die Vorkommen und Veränderungen in der Erdkruste hat (siehe zur Unterscheidung: Neptunismus).

Pneumatolytic (see Appendix VIIb):

A term applied to a supercritical stage, transitional between liquid magmatic and hydrothermal and very closely related to pegmatitic; considered to be responsible for processes of alteration, impregnation, and (contact-) metasomatism; approximate temperature range 400° to 600 °C (about 800 °F to 1,150 °F).

Neumatolítico: Término aplicado a un estado transitorio supercrítico entre el líquido magmático y el hidrotermal y muy estrechamente relacionado con el pegmatítico. A este estado le serían imputados los procesos de alteración, impregnación y metasomatismo (de contacto); la temperatura aproximada varia entre 400° y 600 °C.

Pneumatolytique: Terme qui s'applique à l'état supercritique faisant transition entre l'état magmatique liquide et l'état hydrothermal. L'état pneumatolytique est très proche de l'état pegmatitique. Des processus d'altération, d'imprégnation, de métasomatose (de contact) lui seraient imputables; température approximative de 400° à 600 °C.

Pneumatolytisch: Der Ausdruck wird verwendet für das kritische Stadium im Übergang zwischen der flüssigen magmatischen und hydrothermalen Phase und ist nahezu identisch mit der pegmatitischen Phase. Im pneumatolytischen Stadium finden Umwandlungs- und Imprägnierungsprozesse sowie auch Kontaktmetasomatose statt. Die Temperaturen bewegen sich zwischen 400 und 600 °C.

Pod:

Elongated lentil: A flat elongated ore body, usually of large dimensions. (SCH.)

Pod: Masa en forma de lenteja alargada (cuerpo de mineral alargado y achatado) generalmente de grandes dimensiones.

Lentille allongée: Corps minéralisé plat et allongé, ordinairement de grandes dimensions.
—: Ein langgestreckter linsenförmiger Erzkörper von gewöhnlich großer Ausdehnung.

Primary:

Of rock minerals: those originally present in the rock, not introduced or formed by alteration or metamorphism. Of ore: not enriched or oxidized by supergene processes. (McKinstry)

Primario: En minerales de roca: aquellos originalmente presentes en la roca, no introducidos ni formados por alteración o metamorfismo. De mena: no enriquecidos u oxidados por procesos supergénicos.

Primaire: En ce qui concerne les minéraux des roches: il s'agit des minéraux de première génération, c'est-à-dire qui ne furent pas introduits par altération ou métamorphisme. Pour les minerais: le terme s'applique aux minerais qui ne sont ni oxydés, ni enrichis par des processus supergènes.

Primär: 1. Primäre Mineralien: Jene Mineralien, die ursprünglich im Gestein vorhanden waren und nicht nachträglich durch Umwandlung oder Metamorphose eingeführt oder gebildet wurden. 2. Primäre Erze: Jene Erze, die nicht durch supergene Vorgänge angereichert oder oxydiert sind.

Propylitization:

The deuteric or hydrothermal alteration of andesitic and related rocks into greenstone-like rocks (*propylites*) composed essentially of chlorite, sericite, epidote, carbonates and quartz with disseminated pyrite.

Propilitización: La alteración deutérica o hidrotermal de rocas andesíticas o afines en un tipo de roca (como *greenstone*) llamada propilita, compuesta esencialmente de clorita, sericita, epidota, carbonatos y cuarzo, y en la que es característica la pirita diseminada.

Propylitisation: Altération deutérique ou hydrothermale des andésites et roches connexes en roches vertes (*propylites*) composées essentiellement de chlorite, séricite, épidote, carbonates et quartz, la pyrite disséminée apparaissant caractéristique.

Propylitisierung: Die deuterische oder hydrothermale Umwandlung von andesitischen und diesen verwandten Gesteinen in propylitische Gesteine (Grünsteine); letztere setzen sich im wesentlichen aus Chlorit, Serizit, Epidot, Karbonaten und Quarz mit eingestreutem Pyrit zusammen.

Protore:

Below ore-grade mineralization, usually (as in porphyry copper deposits) underneath a zone of primary ore.

Protomena (protore): Mineralizaciones con una ley no económica, generalmente (como en los depósitos de cobre de tipo porfirítico) debajo de una zona de mineral primario.

Protore: Minéralisation diffuse de teneur inférieure à celle d'exploitabilité, généralement situées (comme dans les gisements de porphyres cuprifères) sous une zone de minerai primaire.

—: Teile einer Vererzung, die nicht abbauwürdig sind und unter der bauwürdigen Zone liegen (z. B. in den Imprägnationslagerstätten).

Pyrometasomatic:

Formed by metasomatic (cf.) processes, usually near magmatic contacts under high temperatures and pressures.

Pirometasomático: Formado por proceso de metasomatismo (véasealli), generalmente cerca de los contactos magmáticos bajo alta temperatura y presión.

Pyrométasomatique: Formé par des processus métasomatiques (voir ce terme), généralement près de contacts magmatiques, c'est-à-dire sous de hautes températures et de fortes pressions.

Pyrometasomatisch: Durch metasomatische (siehe dort) Vorgänge gebildet; gewöhnlich in der Nähe von magmatischen Kontakten unter hohen Temperaturen und Drucken.

Raise

(rise): An opening like a shaft made in the back of a level to reach the level above. (McKinstry) (cf. winge).

Chimenea: Un orificio como un pozo que se hace sobre un nivel para alcanzar el inmediato superior. (cf. winze).

Montage: Ouverture similaire à un puits faite à partir d'un niveau pour atteindre le niveau supérieur. (cf. descenderie).

Aufbruch: Ein Grubenbau ähnlich einem Schacht, der von einer Sohle in die nächsthöhere getrieben wird. (cf. Gesenk).

Recrystallization:

A reorganization of a crystalline aggregate to a compositionally and/or texturally different aggregate; often due to heat and/or stress; often resulting in a general increase in the grain size (cf. *"collection-recrystallization"*).

Recristalización: La reorganización de un agregado cristalino a otro de composición y/o textura diferente; generalmente debido a calor y/o tensión y frecuentemente con un mayor desarrollo del tamaño del grano (ver recristalización de colección).

Recristallisation: Réorganisation d'un agrégat cristallin en une association de composition et (ou) de texture différentes; souvent due aux effets thermiques et (ou) de pression; il en résulte généralement une augmentation de cristallinité. (Voir recristallisation par coalescence).

Rekristallisation: Die Reorganisation eines kristallinen Aggregats zu einem nach Zusammensetzung und/oder Textur verschiedenen Gebilde; häufig verursacht durch Hitze und/oder Spannung und meistens verbunden mit einer generellen Zunahme der Korngröße (siehe Sammelkristallisation).

Reef:

1. (Australia and S. Africa) a lode, vein, or stratiform ore body; a term originally used for gold-bearing quartz veins or (conglomerate) beds; 2. (S. Africa) the barren shales which surround the diamantiferous rocks; 3. In stratigraphy: a large bioherm.

Reef: 1. (Australia y SAfrica) Un filón, veta o mena estratiforme, término originalmente usado para vetas o capas de cuarzo auríferas. 2. (Africa del Sur) Las lutitas estériles ubicadas en los alrededores de las rocas diamantíferas. 3. En estratigrafía: un bioherm grande.

Reef (terme intraduisible en français): 1 — (Australie et Afrique du Sud) Lode, filon ou gisement stratiforme; le sens étant généralement restreint aux formations quartzo-aurifères. 2 — (Afrique du Sud) Shales stériles qui encaissent les roches diamantifères. 3. Un grand bioherm (récif).

Reef: 1. In Australien und Südafrika: Erzgang oder schichtige erzhaltige Lage (z. B. goldführendes Konglomerat). Der Ausdruck wurde ursprünglich für goldführende Quarzgänge und Konglomerate verwendet. 2. In Südafrika: Schiefrige Zonen, die das diamantführende Gestein umgeben. 3. Riff: ein großes Bioherm.

Refractory:

Material with very high melting point, able to resist heat, as in a smelter furnace.

Refractario: Material con un punto de fusión muy alto y capaz de resistir el calor, como en los hornos de fundición.

Réfractaire: Matériaux possédant une température de fusion élevée, capable de résister à la chaleur, comme par exemple dans un four de fonderie.

Feuerfest: Material mit sehr hohem Schmelzpunkt, das großer Hitze ohne Zerstörung widerstehen kann; Verwendung z. B. in Hochöfen.

Regional metamorphism:

Changes of country rock or of a mineral deposit taking place on a regional scale under tectonic pressure and temperature, without any obvious igneous body interacting.

Metamorfismo regional: Cambios de la roca o de un yacimiento mineral en escala regional, bajo presión y temperatura elevadas, pero sin estar necesariamente en conexión con actividad magmática.

Métamorphisme régional: Changement régional d'une roche ou d'un gisement effectué par pression et température élevées, mais généralement sans la présence d'une intrusion.

Regionalmetamorphose: Metamorphe Umwandlung des Gesteins oder einer Lagerstätte; von regionaler Ausdehnung, gewöhnlich über lange Zeitspannen, in der Regel ohne Einfluß magmatischer Gesteine.

Rejuvenation:

1. Of faults: repeated, renewed movement along an old plane. 2. Of mineralizations: renewed supplies of mineral matter or solutions coming in; produce telescoping or at least repeated deposition in veins.

Rejuvenecimiento: 1. De fallas: repetición, reanudación de movimientos a lo largo de un plano antiguo. 2. De mineralización: suministro renovado de materia mineral o de soluciones metalíferas esto conduce a la deposición telescópica o cuando menos a una repetición de deposiciones en las vetas.

Réjuvénation: 1- des failles: mouvements répétés, renouvelés le long d'un plan de cassure ancien. 2- des minéralisations: un nouvel apport de matière minérale ou de solutions métallifères; ceci conduit au téléscopage ou, au moins, à une répétition des dépôts dans les filons.

Rejuvenation: 1. von Störungen: Erneute Bewegung entlang alter Störungsebenen. 2. der Mineralisation: Erneute Zufuhr von Mineralien oder Lösungen; Rejuvenation kann teleskopartige Anreicherungen oder doch wiederholte Ablagerungen verursachen.

Replacement vein:

See *vein*.

Veta de reemplamiento: Ver veta.

Filon de remplacement: Voir filon.

Verdrängungslagerstätte: Siehe Gang.

Residual:

1. Characteristic of, pertaining to, or consisting of residuum. 2. Remaining essentially in place after all but the least soluble constituents have been removed. (AGI)

Residual: 1. Característico de los residuos, perteneciente a ellos, que consiste en ellos. 2. Que permanece en su lugar después que los constituyentes más solubles han sido removidos.

Résiduel: 1 – Caractéristique, comportant ou essentiellement constitué de formations héritées d'un état antérieur. 2 – Ce qui reste en place après que les constituants les plus solubles aient été entraînés.

Residual: 1. Charakteristisch für, bezogen auf, oder bestehend aus Rückständen (Überbleibsel). 2. Material, das an Ort und Stelle verbleibt, nachdem nahezu alle, mindestens aber die löslichen Bestandteile verschwunden sind.

Rock burst:

A sudden and often violent failure of masses of rock in quarries, tunnels and mines. (AGI)

Explosión de roca: Desprendimiento repentino, y con frecuencia violento, de masas en canteras, túneles y minas.

Coup (de toit, de mine . . .): Rupture, souvent soudaine et violente, de masses rocheux dans les carrières, tunnels et mines.

Bergschlag: Ein plötzliches, unvorhergesehenes und oft katastrophales Bersten von Gesteinsmassen in Steinbrüchen, Tunneln und Bergwerken.

Rock-stratigraphic boundaries

Boundaries of rock-stratigraphic units are placed at positions of lithologic change. Boundaries are placed at sharp contacts or may be fixed arbitrarily within zones of gradation. Both vertical and lateral boundaries are based on the lithologic criteria that provide the greatest unity and practical utility. (CSN)

Límites litoestratigráficos: Los límites de unidades litoestratigráficas están situados en los lugares de cambios litológicos. Los límites son dados por contactos bien definidos o pueden ser fijados arbitrariamente dentro de zonas de gradación. Límites verticales y laterales están basados en el criterio litológico que suministra el conjunto mayor y la utilidad práctica.

Limites lithostratigraphiques: Les limites lithostratigraphiques sont basées sur des changements lithologiques, soit nets, soit par convention à l'intérieur d'une zone de changements progressifs. Verticales ou latérales, elles sont choisies sur des critères lithologiques de façon à donner des ensembles qui ménagent à la fois unité et utilité d'emploi.

Gesteinsstratigraphische Grenzen: Grenzen gesteinsstratigraphischer Einheiten liegen an lithologischen Änderungen. Grenzen treten auf an scharfen Kontakten oder können willkürlich in Übergangszonen angenommen werden. Auf Grund lithologischer Kriterien werden sowohl vertikale als auch horizontale Grenzen angenommen in dem Maße, wie sie praktischen Nutzen und größte Vereinheitlichung versprechen.

Rock-stratigraphic unit:

A rock-stratigraphic unit is a subdivision of the rocks in the earth's crust distinguished and delimited on the basis of lithologic characteristics. (CSN.)

Unidad litoestratigráfica: Es una subdivisión de rocas en la corteza terrestre diferenciadas y delimitadas sobre la base de sus características litológicas.

Unité lithostratigraphique: Subdivision de couches rocheuses définie et délimitée par ses caractères pétrographiques.

Gesteinsstratigraphische Einheit: Eine Unterteilung der Gesteinsschichten in der Erdkruste, unterschieden und abgegrenzt nach lithologischen Gesichtspunkten.

Room:

A wide working space in a mine (see *room and pillar*).

Bovedón (room): Espacio amplio en las labores mineras (ver bovedones y pilares).

Chambre: Tout espace de grande dimension dans les travaux miniers. (Voir chambres et piliers.)

Kammer: Ein weiträumiger Grubenbau in einem Bergwerk (siehe Kammerpfeilerbau).

Room and pillar:

See mining methods.

Bovedones y pilares: Ver métodos de minería.

Chambres et piliers: Voir méthodes minières.

Kammerpfeilerbau: Siehe Abbaumethoden.

Rotary drilling:

The most common method of drilling wells. A cutting bit, of varying design for different kinds of rock, is screwed to one or more hollow drill collars to give added weight and these are screwed onto the bottom of the hollow drill pipe, which is rotated from the surface. A stream of drilling mud passes down the drill pipe and the drill collar and out through the bit, from where it is forced back up the hole outside the drill pipe and into pits, where it is again picked up by the pumps and forced back down. Holes over 21,000 feet [7,000 meters] deep have been drilled in this manner. (LEVORSEN)

Sondaje por rotación: El método más común para barrenar pozos. Una corona o broca, de diseños variables para diferentes clases de rocas, es adaptada a una o más barras huecas para darles mayor peso y el conjunto se hace girar desde la superficie. Se hace pasar un flujo de barro a través de barras y corona, desde donde se le fuerza a volver, por fuera de las barras, hasta la superficie, siendo recogido en pozos y volviendo a circular, mediante bombas, en forma continua. De esta manera han sido barrenados huecos de más de 21,000 pies de profundidad.

Sondage rotary: Méthode la plus courante pour foncer des puits. Un outil tranchant, de caracteristiques différentes en fonction des différentes roches à traverser, est vissé à un ou plusieurs colliers − ceci afin d'augmenter le poids − eux-mêmes vissés sur les tubes de forages entraînés depuis la surface dans leur mouvement de rotation. Un courant de boue de forage passe au travers des tubes, des colliers, de l'outil d'où il ressort pour remonter par pression entre les tubes et les parois du trou jusqu'à la surface: là les boues sont recueillies dans des bassins et, après épuration, reprises par des pompes et réinjectées, sous pression, dans le trou. Des forages de plus de 7000 mètres ont été exécutés de cette manière.

Rotary-Bohrverfahren: Die bekannteste Methode zum Niederbringen von Bohrlöchern. Eine Bohrschneide von verschiedener Ausbildung für unterschiedlich beschaffenes Gestein wird auf Hohlbohrstangen, die das notwendige Gewicht geben, aufgeschraubt. Das Stangensystem wird bis zum Bohrlochtiefsten eingelassen und von der Tagesoberfläche aus gedreht. Eine Bohrtrübe wird durch die Bohrstangen nach unten gedrückt, sie tritt aus den Bohrschneiden aus und schwemmt das Bohrklein im Ringraum zwischen Bohrstangen und Bohrlochwandungen wieder nach oben; hier wird die Trübe regeneriert und geht über Pumpen erneut in den Kreislauf. Bohrlöcher mit Tiefen über 7000 m wurden auf diese Weise niedergebracht.

Round:

1. The advance accomplished in a drift or other opening by simultaneously blasting a single set of drilled holes. 2. The set of holes drilled in preparation for the blast. 3. The ore or rock broken by the blast. (McKinstry)

Redondo: 1. El avance que se logra en un socavón u otra abertura mediante un disparo simultáneo de una serie simple de taladros. 2. Serie de taladros en preparación para un disparo. 3. Mena o roca rota por un disparo.

1. **Avancement:** Avance réalisée dans un travail minier, puits ou traçages divers, par un unique programme de trous de mines.

2. **Volée:** Ensemble de trous de mine devant exploser dans le même tir.

3. **Enlevure:** La quantité de roche ou de minerai abattue par un tir.

Abschlag: 1. Der Vortrieb, der in einer Strecke oder irgendeinem anderen bergmännischen Hohlraum beim gleichzeitigen Zünden eines Satzes von Patronen in verschiedenen Bohrlöchern erzielt wird. 2. Der Satz der Löcher, die als Vorbereitung für die Sprengung gebohrt werden. 3. Das Erz oder Gestein, das bei einem Zündgang gelöst wird.

Run:

A term employed in the central U. S. A. for a ribbon-like, layered lead-zinc deposit (in Paleozoic limestone or dolomite) following a certain zone or channel in the plane of stratification. (As thinner and/or weaker zones, those runs are often followed by dipping joint faults transitional to pitches or to breccia zones).

Run: Término empleado en la parte central de EE. UU. para indicar una faja de un depósito estratificado de plomo y zinc (en caliza o dolomita paleozoica) que sigue una cierta zona o canal en el plano de estratificación. Estas zonas tan delgadas y/o débiles frecuentemente van seguidas de fallas transitorias de junturas a hilillos o a zonas de brecha.

Run: Terme utilisé dans les Etats-Unis centrales pour désigner un corps minéralisé plat en forme de ruban; plombo-zincifères, ces minéralisations sont situées dans les roches carbonatées paléozoïques et se développent suivant certaines zônes ou chenaux, dans les plans de stratification. Comme ces runs sont des zônes plus minces et (ou) plus faibles elles coïncident souvent avec des zônes faillées et diaclasées qui passent de manière graduelle aux zônes bréchiques ou aux bretelles.

Run (amer.): In den Zentral-USA verwendete Bezeichnung für bänderartige, schichtige Blei-Zink-Vorkommen (in paläozoischen Kalksteinen oder Dolomiten), die einer bestimmten Schicht oder Kanälen in der stratigraphischen Ebene folgen. (Als dünnere und/oder schwächere Zonen werden diese *Runs* oft begleitet durch steil einfallende Bruchfugen, die übergehen in *Pitches* oder in Brekzien.)

Saddle reefs:

Ore matter occurring in the hinges of folded rock; *inverted saddles* are the opposite, i. e., ore minerals in the troughs of folds.

Filones en montura: Mena que ocupa las crestas de una roca plegada; montura invertida es lo contrario, es decir, una mena que ocupa la concavidad de un plegamiento.

Gîte de charnière anticlinale (ou de crochon anticlinal): Minerai se présentant le long des crêtes de roches plissées; un gîte de crochon synclinal est le contraire, c'est-à-dire formé par des minerais existant dans les parties basses des plis.

Sattel-Reefs: Erzvorkommen an Sätteln von gefalteten Gesteinsschichten. **Verkehrte Sättel** führen ebenfalls häufig Erz (Vorkommen im Muldentiefsten von Faltungen).

Saddle vein:

See *vein*.

Veta en montura: Ver veta.

Filon de charnière anticlinale (ou de crochon anticlinal): Voir filon.

Sattelgang: Siehe Gang.

Salband

(German): Parts of a vein or dike next to the country rock.

Salbanda: Parte de la veta o dique inmediata a la roca de caja.

Eponte: Partie du filon ou du dike avoisinant la roche encaissante.

Salband: Grenzfläche oder -zone zwischen Gang und Nebengestein.

Secondary:

Minerals introduced into the rock or formed by metamorphism or alteration. (McKINSTRY)

Secundario: Minerales introducidos dentro de la roca o formados por metamorfismo o alteración.

Secondaire: Minéraux introduits dans une roche ou formés par métamorphisme ou altération.

Sekundär: Minerale, die Gesteinen zugeführt werden oder durch Metamorphose bzw. Umwandlung entstehen.

Section:

Representation of features such as mine workings or geological features on a vertical (or inclined) plane. A longitudinal section is parallel to the strike of a vein or geologic plane. A cross-section is perpendicular to the strike. (McKINSTRY)

Sección: Representación de rasgos, tales como labores mineras o rasgos geológicos, sobre un plano vertical (o inclinado). Una sección longitudinal es paralela al rumbo de una veta o plano geológico. Una sección transversal es perpendicular al rumbo.

Coupe: Représentation de caractères géométriques, tels que travaux de mines ou éléments géologiques, sur un plan vertical (ou incliné). Une coupe longitudinale est parallèle à la direction d'un filon ou d'un plan géologique. Une section transversale est perpendiculaire à la direction.

Schnitt, (Profil): Die zeichnerische Wiedergabe von Konturen wie z. B. Grubenbauen oder geologischen Erscheinungen in einer vertikalen oder geneigten Ebene. Ein Längsschnitt verläuft parallel, ein Querschnitt senkrecht zum Streichen des Ganges oder der Lagerstätte.

Selvage:

A layered zone of usually soft clayey material, separating the country rock from the ore matter along a fault, joint, fissure or (any other) vein. (Occasionally the contact surface of a vein with the country rock is also simply called *selvage*).

—: Una zona con capas suaves de material arcilloso que separa la mena de la roca estéril a lo largo de una falla, diaclasa, fisura o cualquier otra veta. (Ocasionalmente el contacto de la superficie de una veta con la roca de caja se llama también *selvage*).

Salbande: Zône litée, de matériaux argileux, généralement tendre, séparant la roche encaissante du corps minéralisé, le long d'une faille, d'une diaclase, d'une fissure ou d'un filon. (Quelquefois la surface même de contact d'un filon avec la roche encaissante est également appelée ainsi.)

Salband: Eine dünngeschichtete Zone bestehend aus weichem, lehmigem Material, das in Gängen und an Störungen Erz vom Nebengestein trennt. (Meist ist Salband nur die Begrenzungsfläche, die benachbarte Zone aber der Besteg oder die Gangletten.)

Series

(time-stratigraphic): Series is a time-stratigraphic unit next in rank below a *system*. (SCH.)

Serie (tiempo-estratigráfico): Serie es una unidad de tiempo estratigráfico inmediatamente inferior en categoría a sistema.

Série chronostratigraphique: La série est l'unité chrono-stratigraphique immédiatement inférieure au système.

Serie (zeitstratigraphisch): Eine Einheit, die zeitstratigraphisch eine Untergliederung des Systems bedeutet.

Shaft:

A well-like downward opening in a mine. Usually a shaft starts from the surface; if started from underground workings it is called a *winze*, though if equipped with a hoist it may be called an underground shaft. A shaft may be vertical or inclined. In some districts a flatly inclined opening is called an *incline*, "shaft" being limited to vertical openings. (McKINSTRY)

Pique: Abertura descendente a manera de pozo en una mina. Un pique generalmente comienza en la superficie; si se comienza a partir de las labores subterráneas se llama un *winze*, aunque si está equipado con un elevador podría llamarse un pique subterráneo. Un pique puede ser vertical o inclinado. En algunos distritos una abertura poco inclinada se llama plano inclinado, (*incline* en inglés), reservándose p i q u e para las aberturas verticales.

Puits (de mine): Ouverture de mine dirigée vers le bas. Généralement un puits commence à la surface; s'il est creusé à partir de travaux souterrains, il est appelé descenderie, bien qu'il puisse être appelé bure, s'il est équipé d'un treuil. Un puits peut être vertical ou incliné. Dans certains districts une ouverture peu inclinée est appelée plan incliné (incline en anglais), le terme de puits étant réservé aux ouvertures verticales. N. B. L'acception française des termes utilisés dans la traduction de cette définition est différente et dépend davantage de la façon dont est creusée une ouverture que de ses caractères géométriques. Un *puits* est vertical et foncé par en haut, à partir du jour; au fond, toutes choses égales par ailleurs, il s'agit d'un *bure* (cf. *winze* pro parte). Lorsque l'ouvrage démarré par en haut, que ce soit au jour ou au fond, est incliné par rapport à la verticale, il s'agit d'une *descenderie*. Au contraire, l'ouverture attaquée par en bas, verticale ou inclinée, est un *montage*: il s'agit généralement d'ouvrages à petite section.

Schacht: Ein senkrechter oder stark geneigter Grubenbau, der der Förderung, Bewetterung und Seilfahrt in Bergwerken dient. Gewöhnlich beginnt der Schacht an der Erdoberfläche; Schächte, die Grubenbaue untertage senkrecht verbinden, werden als Blindschächte bezeichnet (Aufbruch oder Gesenk). Tonnlägige Schächte sind geringfügig aus der Vertikalen geneigt. Bei größerer Neigung eines Grubenbaus geht die Bezeichnung Schacht in Förderberg über.

Shear: (see Fig. 3a and 3b):

Mode of failure of a body whereby the portion of the body on one side of a plane or surface slides past the portion on the opposite side. Surface on which shearing has occurred. (McKINSTRY)

Cizalla: Procedimiento de ruptura en el cual la porción de masa situada a un lado de un plano o superficie se desliza sobre el del lado opuesto. Designa también la superficie a lo largo de la cual ha tenido lugar un cizallamiento.

Cisaillement: Mode de rupture dans lequel la portion du corps situé d'un côté d'un plan ou d'une surface glisse le long de la portion située de l'autre côté. *Shear* désigne aussi la surface de cisaillement.

Scherung: Eine Art der Zerstörung, bei der ein Teil eines Körpers entlang einer Bruchfläche gegenüber dem zweiten Teil des Körpers verschoben wird. Die Bruchfläche wird als Scherfläche bezeichnet.

Shear zone:

A layer or slab-like portion of a rock mass traversed by closely spaced surfaces along which shearing has taken place. (McKINSTRY)

Zona de cizallamiento: Porción de una capa o planchón en una masa de roca atravesada por superficies tupidas a lo largo de las cuales ha tenido lugar el cizallamiento.

Zone de cisaillement: Couche ou tranche de roche traversée par des surfaces rapprochées le long desquelles il y a eu un cisaillement.

Scherungszone: Eine geschichtete Gesteinsmasse, die von nahe beieinanderliegenden Bruchflächen, an denen Scherwirkungen stattgefunden haben, durchsetzt ist.

Shear-zone deposit:

An ore deposit formed in a system of interlacing openings parallel to the shearing and mostly of infinitesimal size. After mineralization by filling and/or replacement a wide tabular, massive lode or a lens-like mass of irregular shape is produced. (Sch.)

Depósito en zona de cizallamiento: Depósito mineral formado en el interior de un sistema de entrelazamiento de aberturas paralelas al cizallamiento y en su mayoría de tamaño infinitesimal. Después de la mineralización por relleno y/o reemplazamiento, se produce un depósito mineral amplio tabular, un venero o una lente de forma irregular.

Gîte de zone de cisaillement: Gîte formé à l'interieur d'un système d'ouvertures entrecroisées, parallèles «en grand» à la zone de cisaillement et d'une dimension infinitésimale pour la plupart. Après minéralisation par remplissage et (ou) remplacement il en résulte un lode massif et tabulaire ou une masse lenticulaire de forme irrégulière.

Vererzte Scherzone: Lagerstätte in einer Scherzone verschiedensten Ausmaßes.

Sheeted vein deposit:

See vein.

Depósito de vetas laminadas: Ver veta.

Filon feuilleté: Voir filon.

Tafelförmige, plattige Ganglagerstätte: Siehe Gang.

Shift:

A day's work for a miner (commonly 8 hours). The group of men who work during a given 8-hour period, hence day shift, afternoon shift, night shift, graveyard shift (the shift beginning at or around midnight). (McKinstry)

Turno: Día de trabajo para un minero (generalmente 8 horas). Grupo de hombres que trabajan un período dado de 8 horas, de aquí que haya turno de día, de tarde, de noche y de media noche (turno que comienza a media noche o cerca de esa hora).

Poste: Travail fourni par le mineur en une journée (généralement 8 heures). Groupe d'hommes travaillant pendant une période donnée de 8 heures, d'où les termes de poste de jour, poste d'après-midi, poste de nuit, poste de minuit (en anglais: poste de la nuit et des tombeaux).

Schicht: Die Tagesarbeit eines Bergmanns (gewöhnlich 8 Stunden). Eine Gruppe von Bergleuten, die geschlossen eine achtstündige Arbeitszeit verrichtet (Frühschicht, Mittagsschicht und Nachtschicht).

Shrinkage stoping:

See mining methods.

Testeros con tolvas: Ver métodos de minería.

Chambre-magasin: Voir méthodes minières.

Firstenstoßbau: Siehe Abbaumethoden.

Silicification:

Term which implies introduction of and replacement by silica, but which is often applied without supporting evidence.

Silicificación: Término que entraña introducción o reemplazamiento por sílice, pero que se aplica frecuentemente sin un apoyo de evidencia.

Silicification: Terme impliquant introduction de silice et remplacement par celle-ci, mais qui est souvent utilisé sans preuves.

Verkieselung (Silifizierung): Ein Ausdruck für die Einführung von SiO$_2$ oder Verprängung durch Silikate, oft verwendet ohne sichere Beweise.

Sill

A tabular bed of hard sandstone or of intrusive rock concordant to its layered country rock.

Sill: Cuerpo tabular de arenisca dura o de roca ígnea intrudida paralela a las capas que lo contienen.

Sill: Feuillet de grès dure ou de roche ignée intrusif parallèle aux roches encaissantes.

Lagergang: Plattenförmige, konkordante Einlagerung einer harten Sandsteinschicht oder einer intrusiven Eruptivgesteinsmasse in Schichtverbände.

Skarn:

Rock resulting from contact or regional metamorphism and characterized by calcium-silicates such as garnet and pyroxene to form a tactite.

Skarn: Roca resultante de un metamorfismo de contacto o regional y caracterizada por silicatos de calcio, tales como granate y piroxeno formando una tactita.

Skarn: Roche résultant d'un métamorphisme de contact ou régional caractérisée par des silicates de calcium tels que grenats et pyroxènes formant une tactite.

N. B.: Dans la terminologie française skarn désigne le résultat d'un métamorphisme général de matériaux carbonatés, le¦ terme tactite étant plus spécialement dévolu au faciès corné du métamorphisme de contact des mêmes roches.

Skarn: Gesteine, die durch Kontakt- oder Regionalmetamorphose entstehen und durch Calcium-Silikate wie Granate und Pyroxene gekennzeichnet sind.

Skarn ore:

Ore matter found in, and formed by the same process as the common skarn rocks and minerals (cf. skarn).

Mena de skarn: Mena que se encuentra en el skarn y ha sido formada por el mismo proceso que una roca o minerales comunes de tipo skarn (ver skarn).

Minerai de skarn: Minerai trouvé dans le skarn et formé par le même processus que celui-ci et ses minéraux (voir skarn).

N. B.: L'acception française porte davantage sur le caractère support (sens descriptif pur) que sur l'origine commune.

Skarnerze: Erzbildungen, die durch die gleichen Vorgänge wie die gewöhnlichen Skarngesteine und -minerale entstehen (siehe Skarn).

Slickenside:

Shiny and occasionally grooved fault surface feature.

Espejo de falla: Plano de falla brillante y a veces acanalado.

Miroir de faille: Plan de faille brillant et parfois strié. (En langage minier on utilise aussi le terme de lisses).

Harnisch, Rutschharnisch: Glänzende und manchmal auch gerillte Störungsfläche.

Sludge:

Slurry of rock powder (diamond drilling) or rock cuttings with slurry (churn drilling) made by the drill bit and bailed, or washed to the surface.

Cieno de perforación: Barro, lodo formado de roca pulverizada (perforación diamantina) o de recortes de roca (sondaje a percusión) debidos al barrenaje y bombeadas a la superficie.

Boue (ou *schlamm**): Boue formée de roche pulvérisée (sondage au diamant) ou de débris mélangés à de la boue (sondage au trépan) produits par l'outil et écopée ou entrainée à la surface par un courant.

Bohrschlamm: Trübe mit Gesteinsstaub oder -körnern als festen Bestandteilen, die bei der Arbeit des Meißels auf der Bohrlochsohle entstehen. Die Trübe wird zutage gepumpt und geklärt.

Soft ground:

Parts of mineral deposits which usually can be mined without drilling and blasting of hard rock and in which caving is apt to occur; frequently coincides with upper and/or weathered or altered zones.

Terreno suave: Partes de un depósito mineral que generalmente pueden trabajarse sin perforación o disparo de roca dura y en las cuales pueden producirse con facilidad hundimientos; frecuentemente coincide con las zonas superiores y/o intemperizadas o alteradas del depósito.

—: Parties d'un gisement qui peuvent généralement être exploitées sans perforation ni tir de roche dure et dans lesquelles il peut y avoir foudroyage; elles coïncident fréquemment avec les zones supérieures et (ou) altérées des gisements.

—: Lagerstätten, die gewöhnlich ohne Bohr- und Schießarbeit — jedoch nicht im Tiefbau — hereingewonnen werden. Es handelt sich meistens um oberflächennahe und/ oder verwitterte Vorkommen.

Soil-stratigraphic unit:

A soil-stratigraphic unit is a soil with physical features and stratigraphic relations that permit its consistent recognition and mapping as a stratigraphic unit. (CSN.)

Unidad de suelo-estratigráfico: Es un suelo con rasgos físicos y relaciones estratigráficas que permiten su reconocimiento ininterrumpido y representación en mapa como unidad estratigráfica.

* Terme allemand utilisé en français.

Unité pédostratigraphique: Une unité pédostratigraphique est un sol présentant des caractères physiques et des relations stratigraphiques permettant sa détermination factuelle et sa cartographie en tant qu'unité stratigraphique.

Bodenstratigraphische Einheit: Die Bezeichnung für einen Boden mit bestimmten physikalischen Eigenschaften und stratigraphischen Merkmalen, die eine wiederkehrende Erkennbarkeit und eine Kartierung als stratigraphische Einheit erlauben.

Spud

(in): To commence actual drilling operations on a well. (LEVORSEN)

Spud (in): Comienzo de las operaciones de una perforación en un pozo.

Démarrer un forage d'exploitation: Commencer les véritables opérations de forage sur un puits d'exploitation.

—: Das Ansetzen einer Bohrung.

Spur

A short, irregular branch spreading out from a vein. (SCH.)

Ramal: Brazo corto e irregular que se aparta de una veta.

Apophyse: Branche courte et irrégulière s'écartant d'un filon.

Ast: Ein kurzer, unregelmäßig ausgebildeter Abzweig von einem Erzgang.

Square set:

A set of timbers consisting of vertical and horizontal pieces all meeting at angles of 90 degrees. A system of square sets forms a three-dimensional lattice-work. (MC KINSTRY)

Cuadros: Serie de maderas que consisten en piezas verticales y horizontales que se unen perpendicularmente. Una serie de cuadros de madera formando un enrejado tridimensional.

Cadre trirectangle: Ensemble de bois formé de pièces verticales et horizontales toutes perpendiculaires entre elles. Un système de cadres trirectangles forme un réseau tridimensionnel.

Rahmen, Rahmenbau, Rahmenzimmerung, Geviertzimmerung: Dreidimensionales Ausbaunetz von Holzverstrebungen (Kanthölzer) in vertikaler und horizontaler Richtung, meist in Winkeln von 90°. Ein Rahmenbau kann einfach oder vielfach wiederholt sein.

Square-setting:

(Square-set sto*ping*): See *mining methods*.

Tajeo en cuadros: Ver métodos de minería.

Chambre vide charpentée: Voir méthodes minières.

Blockbau mit Geviertzimmerung, Rahmenbauverfahren: Siehe Abbaumethoden.

Stage:

(time-stratigraphic): Stage is a time-stratigraphic unit next in rank below a *series*. (SCH.)

Etapa (cronoestratigráfico): Unidad de tiempo estratigráfico de categoría inmediatamente inferior a serie.

Étage chronostratigraphique: L'étage est l'unité chronostratigraphique immédiatement inférieure à la série.

Abschnitt (zeitstratigraphisch): Eine Einheit, die zeitstratigraphisch eine Unterteilung der Serie bedeutet.

Step vein:

See *vein*.

Veta en escalón: Ver veta.

Filon en baïonette: Voir filon.

Stufengang: Siehe Gang.

Stock:

An intrusive rock mass smaller than, but similar to a *batholith* (cf.); circular, elliptical, or irregular in cross section, normally with steep sides.

Stock: Masa de roca intrusiva más pequeña pero similar a un *batolito* (ver esta voz); visto en sección transversal es circular, elíptico o irregular, pero normalmente con lados empinados.

Stock: Masse intrusive de roche, plus petite qu'un batholithe mais semblable (voir *batolite*); de section horizontale circulaire, elliptique ou irrégulière, normalement avec des côtés très abrupts.

Stock: Eine intrusive Masse Gesteins, die zwar kleiner als, aber doch ähnlich einem *Batholithen* (siehe dort) ist. Der Stock hat einen kreisförmigen, elliptischen oder auch unregelmäßigen Querschnitt und gewöhnlich steile Seitenbegrenzungen.

Stockwork (see Fig. 2a, 2b; 16):

Stock: A mass of rock irregularly fractured in various directions by short fissures along which mineralization has spread.

Stockwork: Stock − Masa de roca irregularmente fracturada en varias direcciones por pequeñas fisuras a lo largo de las cuales se extiende la mineralización.

Stockwork: Stock − masse de roche, irrégulièrement fracturée dans plusieures directions par des fissures courtes, le long desquelles s'est répandue la minéralisation.
N. B. L'acception française implique une densité de fracturation sensiblement égale dans les trois directions de l'espace.

Stockwerk: Vererztes Gestein, das in vielfacher Richtung schmale Risse aufweist, die wiederum mineralisiert sind.

Stope:

An underground opening from which ore had been or is being extracted. Usually applied to vertical or highly inclined veins.
An *overhand stope* is made by working upward from a level into the ore above. An *underhand stope* is made by working downward beneath a level (McKinstry).

Stope, pique de extracción: Hueco subterráneo del cual ha sido o será extraida la mena. Aplicado en general a vetas verticales o muy inclinadas.

Un «overhand stope» está trabajado desde un nivel determinado hacia el mineral super-yacente, mientras que un "underhand stope" está trabajado desde un nivel determinado hacia el mineral subyacente.

Chantier d'extraction; Dépilage: Une ouverture souterraine de laquelle le minerai a été ou sera extraît. Un dépilage «en montant» se fait à partir d'un niveau dans le minerai situé au-dessus et un dépilage «en descendant» à partir d'un niveau dans le minerai situé en contrebas.

—: Ein Raum in einer Grube aus welchem Erz ausgebracht wurde oder ausgebracht wird. Im allgemeinen auf steilstehende Gänge angewendet.

Ein Übersichbrechen (Stoßbau) bedeutet Ausbringen des Erzes über einer Sohle, während Untersichbrechen (Strossenbau) ein Ausbringen von Erz unterhalb einer Sohle meint. (Verwandte Ausdrücke: Stoß, Ort; siehe daselbst).

Stratigraphic unit:

See: rock-stratigraphic unit, rock-stratigraphic boundaries, time-stratigraphic unit, biostratigraphic unit, and soil-stratigraphic unit.

Unidad estratigráfica: Ver unidad litoestratigráfica, unidad cronoestratigráfica y unidad suelo-estratigráfica.

Unité stratigraphique: Voir unité lithostratigraphique, chronostratigraphique, bio-stratigraphique, pédostratigraphique.

Stratigraphische Einheit: Siehe gesteinsstratigraphische Einheit, zeitstratigraphische Einheit, biostratigraphische Einheit und bodenstratigraphische Einheit.

Streaks:

Irregular, generally elongated bands, layers or lentils of massive or disseminated ore (compare pattern 10, Appendix VIb).

—: Bandas, capas o lentes irregulares de mineral diseminado o masivo (ver el rasgo 10 del apéndice VIb).

—: Lentilles, bandes ou couches irrégulières, généralement aplaties de minerai disséminé ou massif (voir fig. 10, Appendice VIb).

—: Streifenförmige, schlierige Erzverteilungen (vergleiche Schema 10, Appendix VIb).

Strike:

The bearing of a horizontal line in the plane of a bed, vein, fault, etc. (McKinstry).

Rumbo: Dirección de una línea horizontal en el plano de una capa, veta, falla, etc.

Direction: Direction de l'horizontale dans le plan d'une couche, d'un filon, d'une faille, etc.

Streichen: Die Richtung einer Horizontallinie in der Ebene einer Schicht, eines Ganges, einer Störung u. a.

Stringer (see Fig. 2a, 2b; 16):

A very thin veinlet or string-like occurence of ore matter.

Vetilla: Mena pequeña y estrecha o mineralización filiforme.

Veinule: Filonnet très mince ou minéralisation filiforme.

—: Ein sehr unscheinbarer, fadenähnlicher Erzgang.

Stringer lead (see Fig. 2a, 2b; 16):

Stringer lode: A shattered zone cemented by a network of small nonpersistent veins. (SCH.)

Vetilla guía: Vetilla de un criadero. Zona muy fracturada cementada por una red de pequeñas vetillas no persistentes.

—: Zone fracturée cimentée par un réseau de filonnets discontinus.

—: Eine zerrüttete Gesteinszone durchsetzt von einem Netzwerk kleiner Erzadern.

Stromatites (see App. VIb):

Rocks, the fabric of which forms a system of layers, strata, or bands.

Estromatitas: Fábrica de roca que forma un sistema de capas, estratos o bandas.

Stromatite (peu employé): Roche, dont la fabrique forme un système de couches, de strates ou de lits.

Stromatitische Gesteine weisen ein schichtiges, lagiges, gebändertes oder schlieriges Gefüge auf.

Stull:

A platform (*stull-covering*) laid on timbers (*stull-pieces*) braced across a working from side to side, to support a workman or to carry ore or waste. (RAYMOND, via FAY) However, the term is commonly applied to the timbers (*stull-pieces*) which extend across a stope from hanging wall to foot wall. (MCKINSTRY)

Andamio: Plataforma (andamio-cubierta) que descansa sobre maderos (andamiosoporte) y que sirve para sostener a un trabajador o para soportar mena o desmonte. Sin embargo el término se aplica por lo común a los estemples, los cuales se extienden a través de un tajo desde el techo al piso.

Plateforme: Une plateforme (*stull-covering*) placée sur des buttes (*stull-pieces*) rivées entre les deux parements d'un chantier pour supporter un travailleur ou porter du minerai ou des stériles. Cependant (en anglais) *stull* s'applique couramment aux buttes traversant une chambre de toit au mur.

Arbeitsbühne: Ein Gerüst, das in einem bergmännischen Hohlraum aus Hölzern (*stull-pieces*) errichtet wird, um für Bergleute eine Arbeitsplattform zu schaffen oder um Erz oder Gestein im Zuge des Abbaus vorübergehend aufzunehmen. Im englischen wird der Ausdruck *stull-pieces* für solche Hölzer verwendet, die in einem Abbau das Liegende gegen das Hangende abspreizen.

Sublevel:

An intermediate level opened a short distance below a main level. (FAY)

Sub-nivel: Nivel intermedio, abierto a corta distancia debajo del nivel principal.

Sous-niveau: Niveau intermédiaire, ouvert à peu de distance sous un niveau principal.

Zwischensohle: Eine Sohle, die zwischen den Hauptsohlen eines Bergwerks ausgefahren wird und der unmittelbaren Lösung des zu gewinnenden Minerals dient.

Sublevel stoping:

See *mining methods.*

Tajeo en subniveles: Ver métodos de minería.

Chambre-magasin avec front vertical rabattant sur galerie horizontale préalable: Voir méthodes minières.

Teilsohlenbruchbau, Teilsohlenbau: Siehe Abbaumethoden.

Sub-outcrop:

Intersection of a vein (or other structural feature) with an unconformity. What would be the outcrop if the overlying unconformable formation were removed. (McKINSTRY).

Subafloramiento: Intersección de una veta (o de otro elemento estructural) con una discordancia, lo cual constituiría un afloramiento si la formación discordante estuviese removida.

Pseudo-affleurement: Intersection d'un filon (ou de tout autre élément structural) avec une discordance, ce qui serait un affleurement si la formation discordante susjacente était enlevée.

Pseudoausbiß: Die Unterbrechung eines Ganges (oder einer anderen Gesteinsstruktur) durch eine diskordante Überlagerung. Nach Wegräumen der Überlagerung entsteht ein echter Ausbiß.

Subvolcanic rocks and mineral deposits (see Appendix VII b):

Formed at shallow depth normally with connections to volcanic vents.

Rocas subvolcánicas y depósitos minerales: Formados a poca profundidad, normalmente en conexión con respiraderos volcánicos.

Subvolcaniques, roches et gîtes: Formés à faible profondeur, normalement liés à des cheminées volcaniques.

Subvulkanische Gesteine und Lagerstätten: Gebildet in geringen Teufen, gewöhnlich in Verbindung mit vulkanischen Ausbrüchen.

Supergene:

Generated from above. Refers to the effects (usually oxidation and secondary sulphide enrichment) produced by descending groundwater. Cf. *Hypogene.* (McKINSTRY)

Supergénico: Generado desde arriba. Se refiere a los efectos (generalmente oxidación y enriquecimiento secundario de sulfuros) producidos por aguas subterráneas descendentes. Ver hipogeno.

Supergène: Né de la surface. Réfère aux effets (habituellement l'oxydation et l'enrichissement secondaire des sulfures) produits par les eaux souterraines descendantes. Voir hypogène.

Supergen: Entstanden von oben her; nimmt Bezug auf Wirkungen des Grundwassers, z. B. Oxydation und sekundäre Sulfidanreicherung. Siehe hypogen.

Supracrustal:

Rock, ore and any process taking place on the surface of the earth's crust (sedimentary or volcanic); to be used in place of *exogenetic* (opp. *infra- or intracrustal*).

Supracortical: Roca, mena o cualquier proceso que tiene lugar sobre la superficie de la corteza terrestre (p. ej. sedimentario o volcánico); debe usarse en vez de exogenético (opuesto: infracortical o intracortical).

Supracrustal: Roche, minerai ou n'importe quel processus se produisant à la surface de l'écorce terrestre (p.ex. sédimentaires ou volcaniques); à utiliser en synonyme d'exogène (s'oppose à infra- ou intracrustal).

Suprakrustal: Gesteine, Erze oder Vorgänge, die an der Erdoberfläche liegen oder stattfinden wie Sedimentation und vulkanische Auswirkungen; kann auch anstelle von exogen verwendet werden (Gegenteil: infra- oder intrakrustal).

Swell:

A local enlargement or thickening of an ore vein, an ore bed or a mineral zone (opp. *pinch*). (Sch.)

—: Ensanchamiento o engrosamiento local de una veta mineral, una capa mineral o una zona mineral (opuesto: estrechamiento).

Renflement: Agrandissement ou épaississement local d'un filon, d'un lit ou d'une zone minéralisée (s'oppose à pincement ou étranglement).

Stauchung: Eine örtlich begrenzte Vergrößerung oder Verdickung eines Erzganges oder einer geschichteten Lagerstätte (Gegenteil: *pinch*).

Synchronous:

Processes taking place contemporaneously, or mineral matter forming at the same time (cf. synchronous highs).

Sincrónico: Proceso que tiene lugar contemporáneamente, o materia mineral que se forma al mismo tiempo (ver altitud sincrónica).

Synchrone: Processus contemporain ou matiére minèrale formé en même temps (voir *synchronous highs*).

Synchron: Vorgänge, die gleichzeitig stattfinden oder Minerale, die zur gleichen Zeit gebildet wurden (siehe *synchronous highs*).

Synchronous highs (see Fig. 15a, b):

Sedimentary or topographic features which were formed or are being formed contemporaneously with a certain horizon or unit (may represent ore mineral or oil habitat).

Eminencia o cúpula sincrónica: Caracteres sedimentarios o topográficos que se forman

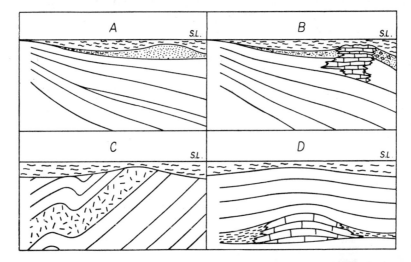

Fig. 15a. Diagram showing synchronous bottom highs of depositional, erosional, and inherited origin. *A:* Sand bar. *B:* Biotherm. *C:* Drowned erosional remnant. *D:* High due to differential compaction over buried reef. S. L. indicates sea-level. (After SCHOLTEN).

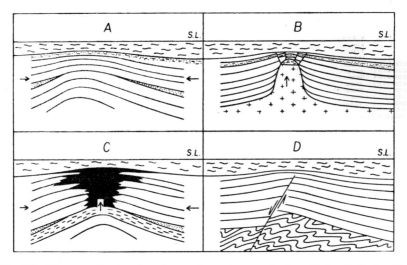

Fig. 15b. Diagram showing four types of synchronous bottom highs of diastrophic origin ("growing structures"). *A:* Early compressional anticline. *B:* Salt core structure. *C:* Mud core structure with intermittent surface extrusion. *D:* Tilted fault block. *S. L.* indicates sea-level. (After SCHOLTEN).

contemporáneamente a un cierto horizonte o unidad (puede representar un ambiente para mena o petróleo).

—: Elévations sédimentaires ou topographiques formés au même moment qu'une unité ou un horizon déterminé (peut représenter des minerais ou des réservoirs).

—: Paläographische, ursprüngliche Erhebung, die sich gleichzeitig mit einer bestimmten Schicht bildete (und Bildungsort einer Erz- oder Öllagerstätte sein kann).

Syngenetic

(with regard to rocks and mineral deposits): Syngenetic are those rocks, minerals, mineral deposits, textures, structures, processes, etc., which were formed contemporaneously with the enclosing rock(s), as contrasted to *epigenetic* rocks, minerals, etc., which are of later date than the enclosing rock(s).

Singenético (con relación a rocas y depósitos minerales): Singenéticos son aquellas rocas, depósitos minerales, texturas, estructuras, procesos, etc., que se forman contemporáneamente a la(s) roca(s) de caja, a diferencia de las rocas, minerales, etc., epigenéticos, que se forman posteriormente a las rocas de caja.

Syngénétique (s'applique aux roches et aux gîtes): Syngénétiques on appelle les roches, minéraux, gîtes, textures, structures, processus formés en même temps que les roches qui les enferment, par opposition aux roches, minéraux, etc. épigénétiques, qui sont plus tardifs que les roches encaissantes.

Syngenetisch (bezogen auf Gesteine und Minerallagerstätten): Syngenetisch sind Gesteine, Minerallagerstätten, Texturen, Strukturen etc., die gleichzeitig mit dem umgebenden Gestein gebildet wurden; im Gegensatz zu epigenetischen Gesteinen, Mineralen etc. die später als das umgebende Gestein entstanden sind.

System

(time-stratigraphic): The system is a time-rock unit usually larger than a *series*, smaller than a *group*, and deposited during a *period*.

Sistema (cronoestratigráfico): Unidad cronolitológica generalmente más grande que una serie y más pequeña que un grupo, depositada durante un período.

Système chronostratigraphique: Le système est l'unité chronolithologique générale-ment superieur á la série et inferieur au groupe et déposé deposíee au cours d'une période.

System (zeitstratigraphisch): Eine Einheit, die zeitstratigraphisch umfassender als eine Serie und enger begrenzt als eine Gruppe ist.

Tailings:

That product of concentration processes which has been separated from the ore minerals, and is deposited as waste into tailing ponds or pumped back into old mine workings (opp.: *concentrates*).

Colas: Producto del proceso de concentración que ha sido separado de los minerales valiosos y es depositado como desmonte en charcas o bombeado de nuevo a las labores mineras antiguas (opuesto: concentrado).

Stériles (ou *Tailings*): Le produit des procédés de concentration qui a été séparé des minerais métalliques et qui est déposé comme déchet dans des bassins ou pompé dans les vieux travaux de mine (Contraire: concentrés).

102

Abgänge: Die beim Aufbereitungsprozeß von Erz, Salz und Kohle anfallenden wertlosen Nebenprodukte, die als Schlamm oder Berge in Teiche oder alte Grubenbaue transportiert werden (Gegenteil: Konzentrat).

Talus:

Streaks, tongues or irregular masses of rock fragments on slopes or at their foot; partly consolidated or loose; may contain eluvial deposits.

Talud: Pedazos, lenguas o fragmentos irregulares de rocas sobre las faldas o en la base de los cerros, parcialmente consolidados o sueltos; pueden contener un depósito eluvial.

Eboulis: Trainées, langues ou masses irrégulières de fragments de roches sur des pentes, ou à leur pied; partiellement consolidé ou meuble; peut contenir des gîtes éluviaux.

Gehängeschutt: Züge, Zungen oder unregelmäßige Massen von Fall- und Rutsch-Schutt an Hängen und am Hangfuß; locker oder teilweise verfestigt; gelegentlich mit eluvialen Seifen.

Telemagmatic deposits (see Appendix VII b):

Formed outside, far away from, and usually with no proved connection with any known intrusions.

Depósitos telemagmáticos: Formados fuera y muy lejos de una intrusión conocida y generalmente sin relación probada con ella.

Gîte télémagmatique: Formé à l'extérieur et loin d'une intrusion connue, et généralement sans relation démontrée avec une telle intrusion.

Telemagmatische Lagerstätten: Lagerstätten, die weit entfernt und gewöhnlich ohne nachweisbare Verbindung mit einer Intrusion gebildet wurden.

Telethermal:

Zone or environment of ore deposition characterized by lesser intensity than *epithermal* and in general by remoteness from an igneous source. (McKinstry)

Teletermal: Zona o medio de deposición mineral, caracterizada por una menor intensidad que el epitermal y en general por su remota distancia a una fuente ígnea.

Téléthermal: Zone ou environnement de minéralisation caractérisé par une intensité inférieure à l'épithermale et, en général, par l'éloignement d'une source ignée.

Telethermal: Der Teil einer Erzlagerstätte, der durch eine geringere Temperatur als im epithermalen Bereich und durch eine größere Entfernung vom magmatischen Herd gekennzeichnet ist.

Tenor:

The metal content or grade of an ore, or a smelter product, given in percents, ounces, etc.

Tenor: Contenido de metal o ley de una mena o de un producto de fundición; se da en porcentajes, onzas, etc.

Teneur: Contenu en métal d'un minerai ou d'un produit de fonderie donné en pourcentage, unités de poids, etc.

Gehalt: Der Metallgehalt im Erz oder in einem Verhüttungsprodukt, angegeben in Prozent oder in Gewichtseinheiten.

Time-stratigraphic(Chronostratigraphic) Units (see Appendix III):
A time-stratigraphic unit is a subdivision of rocks considered solely as the record of
a specific interval of geologic time. (CSN.) The ranks of time-stratigraphic units are,
system, series, and *stage* (compare geologic-time units).

Unidades de tiempo estratígráfico (cronoestratigráficas): Una unidad de tiempo
estratigráfico es una subdivisión de rocas considerada únicamente como la historia
de un intervalo específico de tiempo geológico. Las categorías de unidades de tiempo
estratigráfico son: sistema, serie y etapa (ver unidades de tiempo geológico).

Unités chronostratigraphiques: L'unité chronostratigraphique est une subdivision
basée sur les roches considérées seulement en tant qu'enregistrement des phénomènes
qui se sont développés au cours d'un intervalle de temps géologique donné. La hiér-
archie des unités chronostratigraphiques est la suivante: système, série et étage (com-
parer avec les unités géochronologiques).

Zeitstratigraphische Einheit: Ein Abschnitt in der Gesteinsgliederung, der ein bestimm-
tes Intervall der geologischen Zeit bezeichnet. Die einzelnen Stufen der zeitstrati-
graphischen Einheiten sind System, Serie und Abschnitt (siehe geologische Zeitein-
heiten).

Trough vein:

See *vein*.

Veta en artesa: Ver veta.

Filon de charnière synclinale (ou de crochon synclinal): Voir filon.

Muldengang: Siehe Gang.

Tubing

The pipe through which oil, or gas, or both are brought from the reservoir to the surface. It is generally placed ("run") inside the *casing*, and is 2–4 inches in diameter.
(LEVORSEN)

Tubería: Tubo a través del cual petróleo, gas o ambos son conducidos desde el reservorio a la superficie. Se coloca generalmente dentro de la tubería de revestimiento;
es de dos a cuatro pulgadas de diámetro.

Tubage d'exploitation: Tuyau à travers lequel le pétrole, ou le gaz, ou tous les deux,
arrivent du réservoir à la surface. Il est, en général, placé à l'intérieur du tubage de
soutènement et a de 5 à 10 cm de diamètre.

Verrohrung: Die Rohre, durch die Gas, Öl oder beides vom natürlichen Reservoir bis
zur Erdoberfläche geleitet wird. Bei einem Durchmesser von 5–10 cm ist sie gewöhnlich
in Futterrohren verlegt.

Unit

(as used in connection with *tenor* or price of a metal): The amount of metal (or oxide
or other component) contained in a ton of 1% ore; hence: short ton unit = 20 lb;
long-ton unit = 22.4 lb; metric ton unit = 10 kg (22.04 lb.). Example: if the value of
tungsten ore is $ 20 per short ton unit of WO_3, a ton of 60% ore (since it contains
60 units) is worth $ 1200. (McKINSTRY)

Unidad (tal como se usa en conexión con el contenido o precio de un metal): la cantidad de metal (óxido u otro componente) contenida en una tonelada de 1% de mena;
de ahí que: unidad de tonelada corta = 20 lbs; unidad de tonelada larga = 24.4 lbs;

unidad de tonelada métrica = 10 kg (22.04). Ejemplo: si el valor del mineral de tungs-
teno es de $ 20 por unidad de tonelada corta de WO_3, una tonelada de mineral a 60%
(dado que contiene 60 unidades) vale $ 1200.

Point (employé à propos de la teneur ou du prix d'un métal): Quantité de métal (ou
d'oxyde ou de n'importe quel autre composé) contenu dans une tonne de minerai à
1%. Par conséquent: point de short ton = 20 lb; point de long ton = 22.4 lb; point
de tonne métrique = 10 kg (22.04 lb). Exemple: Si le minerai de tungstène vaut $ 20
le point de short ton de WO_3, une tonne de minerai à 60%, puisqu'elle contient 60
points, vaut $ 1200.

Einheit (Bezogen auf Gehalte und Preise von Metallen): Die Menge an Reinmetall
(oder Oxyd oder anderen Komponenten), die in einer Tonne einprozentigen Erzes enthalten ist. Daher: Einheit der short ton = 20 englische Pfund; Einheit der long ton =
22,4 englische Pfund; Einheit der metrischen Tonne = 10 kg (22,04 englische Pfund).
Beispiel: Bei einem Wert von 20 Dollar für eine short ton-Einheit von WO_3 (Wolframit)
liegt der Wert einer Tonne 60prozentigen Erzes bei 1200 Dollar.

Vadose water:

Water below the earth's surface, in the zone of aeration, but above the water table.

Agua vadosa: Agua subterránea situada en la zona de aeración de la tierra, pero por
encima del nivel freático.

Eau vadose: L'eau souterraine dans la zone aérée de la terre, mais au-dessus du
niveau phréatique.

Vadoses Wasser: Wasser unterhalb der Erdoberfläche, in der Belüftungszone, aber
oberhalb des Grundwasserspiegels.

Vein:

In non-technical mining language the term vein is used for any ore or non-ore material
which exhibits a tabular or sheet like nature, even if this mass is a stratigraphic bed;
for the latter case the semi-technical terms *bed vein* or *blanket vein* are occasionally
used. Such usage is not correct if the term vein implies a cross-cutting body.

Banded vein, Ribbon vein: A crustified vein made up of layers of different minerals,
parallel to the walls (SCH.).

Brecciated vein: A fissure filled with fragments of country rock or earlier generations
of ore matter, in the interstices of which later vein matter is deposited.

Chambered vein: Chambered lode: A vein or lode whose walls are irregular and
brecciated, particularly the hanging wall (partly synon. with the term *stockwork*).

Composite vein: See *lode*.

Contact vein: A vein following the contact between two formations; the contact
is usually caused by faulting in the plane of the fissure. (Note: Not to be confused with
a *contact-metamorphic deposit*). (SCH.)

Countervein; Counterlode: Crossvein or crossveins, running at an angle across
the main ore vein system or lode.

Crustified vein: A vein which has been filled with a succession of crusts of ore and
gangue material. (SCH.)

Fissure vein: A fissure in the earth's crust filled with minerals. (RAYMOND via FAY) As it is now recognized that replacement lateral secretion or) simple filling has played an important role in the formation of most veins, the term fissure vein has lost much of its meaning.

Gash vein: Designating vertical mineralized joint cracks of moderate extent particularly in limestone, usually restricted to one formation.

Ladder vein; Ladder lode: More or less regularly spaced, short, transverse fracture fillings sometimes occuring in dykes. (SCH.)

Lenticular vein; Dilation vein: Thick lenses in schists, thought by some to be caused by the bulging or dilation of schistose rocks due to pressure transmitted by mineralizing solutions. Others would not call them veins but rather only lenses, and consider them to be original sedimentary lenses or lenticular concentrations from the sourrounding rock, by short-range lateral or vertical migrations, i. e., concentrations during folding and metamorphism.

Linked veins: A system of individual more or less parallel veins linked together by cross veinlets. (SCH.)

Main vein: Master vein; major vein; trunk vein.

Network of veins: A not too irregular system of veins striking in different directions.

Replacement vein; Replacement lode: Mineralized fissure (or a system of subparallel fissures) along which, in addition to the fissure filling itself, the walls have to a certain extent been replaced by ore or mineral substance. (SCH.)

Saddle vein; Saddle reef (Australia): Saddle-shaped ore deposit formed between or in sedimentary beds in the crests of, mostly sharply folded, anticlinal structures. (SCH.)

Sheeted vein deposit: An ore deposit occupying a group of closely spaced, distinctly parallel fractures or fissures, separated by narrow plates of country rock. After metallization a composite vein or a replacement lode results, often characterized by a banded structure. (SCH.)

Step vein: A vein alternately cutting through the strata of country rock and running parallel with them. (SCH.)

Trough vein; Trough reef (Australia): Trough-shaped ore deposit formed between or in sedimentary beds in the troughs of, mostly sharply folded, synclincal structures. (SCH.)

Veta: En lenguaje minero no técnico se usa para cualquier material mineral económico (mena) o no económico, los cuales muestran una naturaleza tabular o laminar, incluso si esta masa es una capa estratigráfica; en este último caso el término semitécnico capa-veta o veta manteada se usa muy poco. Estas denominaciones deben ser evitadas si la naturaleza transversal es establecida.

Veta bandeada: Veta en listones − veta costrificada formada de capas de diferentes minerales, paralelas a las cajas.

Veta brechada: Fisura rellena con fragmentos de roca encajante, o de las primeras generaciones de materia mineral en cuyos intersticios se deposita el material de la última generación.

Veta ensanchada, veta cámara: Venero ensanchado − Veta o venero cuyas paredes son irregulares y brechadas, particularmente las paredes superiores; es en cierto modo sinónimo de stockwork.

Veta compuesta: Ver venero.

Veta de contacto: Veta que sigue el contacto entre dos formaciones; el contacto es generalmente causado por fallamiento en el plano de fisura. (Nota: no confundirla con depósito de contacto metamórfico).

Contraveta; contrafilón: Veta o vetas contrarias que corren en un ángulo transversal al sistema principal de vetas o filones de minerales.

Veta costrificada: Veta que ha sido rellenada con una sucesión de costras de mena y ganga.

Veta de fisura: Fisura de la corteza terrestre rellenada con mineral. Tal como se reconoce en la actualidad, el término veta de fisura ha perdido mucho de su significado, debido a que el reemplazamiento, la secrecion lateral y un simple relleno han jugado un papel importante en la formación de la mayoría de las vetas.

Veta en cuña: Designa una juntura vertical de moderada extensión mineralizada, particularmente en caliza y generalmente restringida a una formación.

Veta escalonada: Filón de fracturas transversales cortas, más o menos regularmente esparcidas y presentándose a veces en diques.

Veta lenticular: Dilatación de veta — lente achatado en esquistos, algunos creen que se debe al combamiento o dilatación de rocas esquistosas debido a presiones transmitidas por soluciones mineralizantes. Otros no les llamarían vetas sino simplemente lentes, y las considerarían lentes sedimentarias originales o concentraciones lenticulares procedentes de las rocas circunvecinas, debidas a un corte recorrido por migraciones laterales o verticales; es decir concentraciones durante plegamiento o metamorfismo.

Vetas eslabonadas: Sistema de vetas individuales más o menos paralelas, todas enlazadas por vetillas transversales.

Veta principal: Veta directriz, veta mayor, veta troncal.

Vetas en mallas: Sistema de vetas no muy irregulares, con rumbos en diferentes direcciones.

Veta de reemplazamiento; venero de reemplazamiento — fisura mineralizada (con un sistema de fisuras subparalelas) a lo largo dela cual, además del relleno mismo de la fisura, las paredes han sido en cierta extensión reemplazadas por mena o sustancia mineral.

Saddle vein; saddle reef (Australia): Depósito mineral en forma de montura formado entre (o en) las crestas de capas sedimentarias, en su mayoría estructuras anticlinales fuertemente plegadas.

Depósito de vetas tabulares: Depósito mineral que ocupa un grupo de fracturas muy juntas o fisuras estrechamente separadas y claramente paralelas, separadas por láminas delgadas de roca de caja. Después de la mineralización, resulta una veta compuesta o un venero de reemplazamiento, a menudo caracterizado por una estructura bandeada.

Veta en escalón: Veta que corta alternadamente el estrato de la roca encajonante y se dirige paralelamente a ella.

Veta en artesa: Filón en forma de cubeta — depósito mineral en forma de artesa, formado entre (o dentro de) las despresiones de las capas sedimentarias, en su mayoría estructuras sinclinales fuertemente plegadas (Australia).

Filon: Dans le langage minier non spécialisé, s'applique à n'importe quel matériel, qu'il soit minerai ou non, s'il présente une forme tabulaire ou aplatie, même si cette masse est un lit stratigraphique; dans ce dernier cas le terme semi-technique de filon-couche est quelquefois employé. Cette pratique est à éviter si la nature scéante est une épithète typique du terme filon.

Filon rubané: Filon crustifié constitué de couches de différents minéraux parallèles aux murs.

Filon bréchoïde: Fissure remplie avec des fragments de la roche encaissante ou d'une génération antérieure de minerai, dans les interstices desquels s'est déposé ultérieurement une génération de minerai plus récente.

Filon en chambre: Filon ou lode dont les murs sont irréguliers et bréchiques au toit (en partie synonyme de stockwork).

Filon composé: Voir *lode*.

Filon de contact: Filon longeant le contact entre deux formations; le contact est généralement provoqué par rupture suivant le plan de fracture. N. B.: ne pas confondre avec les gîtes de métamorphisme de contact.

Croiseur (ou filon croiseur): Filon séquent recoupant obliquement le système filonien principal.

Filon crustifié: Filon qui a été rempli par une succession de crustifications de minerai de gangue.

Filon de fissure: Fissure de l'écorce terrestre remplie de minéraux. Comme il est reconnu maintenant que le remplacement, la sécrétion lateral ou simple remplissage a joué un rôle important dans la formation de la plupart des filons, le terme de filon de fissure a perdu beaucoup de sa signification.

—: Désigne des joints minéralisés verticaux de faible extension, particulièrement dans les calcaires et habituellement limités à une seule formation.

Filon en échelle: Remplissage de fractures transversales courtes, plus ou moins régulièrement espacées, se présentant quelquefois dans des dikes.

Filon lenticulaire: Grosses lentilles dans des schistes, imaginées par les uns comme résultats d'un renflement ou d'une dilatation de roches schisteuses dues aux pressions transmises par des solutions minéralisantes. D'autres auteurs les appeleraient plutôt lentilles et les considereraient comme lentilles sédimentaires originales ou comme concentrations lenticulaires à partir de la roche environnante, par des migrations latérales ou verticales de faible amplitude au cours du plissement et du métamorphisme.

Filons anastomosés: Système de filons individuels, plus ou moins parallèles, reliés par des filonnets séquents.

Filon principal.

Système filonien: Ensemble assez régulier de filons à directions différentes.

Filon de remplacement: Fissure minéralisée (ou système de fissures subparallèles), le long de laquelle, en plus de remplissage proprement dit, les parements ont, dans une certaine mesure, été remplacés par du minerai ou des substances minérales.

Filon (gîte) de crochon anticlinal (ou de charnière anticlinale): Gîte en forme de selle mise en place entre ou dans les lits sédimentaires à la crête de structures anticlinales, le plus souvent intensément plissées (Australie).

Filons feuilletés: Gîte occupant un groupe de fractures ou de fissures, très rapprochées et nettement parallèles, séparées par d'étroites plaques de roche encaissante.

Après minéralisation il en résulte un filon complexe ou un lode de remplacement, souvent caractérisé par une structure rubanée.

Filon en baïonnette: Filon coupant alternativement les strates de la roche encaissante et courant parallèlement à elles.

Filon (gîte) de crochon synclinal (ou de charnière synclinale). Gîte en forme de cuvette mis en place entre ou dans les lits sédimentaires dans le creux de structures synclinales, le plus souvent intensément plissées (Australie).

Gang (seltener: Ader): Im allgemeinen bergmännischen Sprachgebrauch wird als Gang jedes Erz oder Mineralvorkommen bezeichnet, das eine tafel- oder schichtartige Form hat, auch wenn es sich um eine stratigraphische Einheit handelt. Für den letzteren Fall werden in der englischen Sprache die Bezeichnungen *bed vein* oder *blanket vein* verwendet, jedoch sollten diese Ausdrücke vermieden werden, wenn ein Querschnitt durch die Lagerstätte typische Hinweise für die Definition Gang zeigt. (Vgl. engl. dike).

Bändergang: Bestehend aus Lagen verschiedener Minerale, parallel zu den seitlichen Begrenzungen des Ganges. (syn.: gebänderter Gang; Banderz meist nur für schichtgebunde Erze gebraucht).

Brekziengang: Eine Kluft gefüllt mit Bruchstücken aus Nebengestein oder Erz früherer Entstehung, wobei die Zwischenräume von dem eigentlichen Gangmaterial ausgefüllt sind.

Kammergang: Ein Vorkommen, dessen seitliche Begrenzungen unregelmäßig und brekzienartig ausgebildet sind – besonders das Hangende (teilweise synonym mit dem Ausdruck Stockwerk).

Kontaktgang: Eine Gangfüllung die der Grenzfläche zwischen zwei Formationen folgt. Die Grenzfläche ist gewöhnlich verursacht durch Störungsbewegungen (nicht zu verwechseln mit einer kontaktmetamorphen Lagerstätte).

Quergang: Er verläuft winklig zur Richtung des Hauptganges.

Spaltengang: Ein Spalt in der Erdkruste, der mit Mineralien gefüllt ist. Seit erkannt wurde, daß Verdrängung und einfache Füllung in gleicher Weise eine bedeutende Rolle bei der Bildung der meisten Erzgänge spielen, hat der Ausdruck Spaltengang viel von seiner Bedeutung verloren.

–: Bezeichnet senkrecht stehende mineralisierte Gangfüllungen von geringer Ausdehnung, meist mit klaffenden Öffnungen, vorwiegend in Kalkgestein. Er ist meistens auf eine Formation beschränkt.

Leitergang: Quer verlaufende Spaltenfüllungen, die in mehr oder weniger gleichmäßigen, kurzen Abständen angeordnet sind; treten manchmal auch in Intrusivgängen auf.

Linsengang: Dicke linsenartige Vorkommen in Schiefergestein. Von einigen Geologen wird angenommen, daß dieser Gangtyp durch Ausbauchung und Dehnung des schiefrigen Gesteins – verursacht durch hohen Druck von mineralführenden Lösungen – entsteht. Andere bezeichnen diese Art des Vorkommens einfach als Linse und vermuten, daß es sich um ursprünglich sedimentäre Lagerstätten oder linsenförmige Konzentrationen aus dem benachbarten Gestein gehandelt hat, z. B. Konzentrationen im Zuge einer Faltung und Metamorphose.

Kettengang: Ein System von einzelnen, parallel laufenden Gängen, die durch kleine Quergänge miteinander verbunden sind.

Hauptgang: Bei einem System von Gängen jener, der die Hauptstreichrichtung angibt.

Netz von Gängen: Ein System von Gängen mit verschiedenen Streichrichtungen.

Verdrängungsgang: Eine mineralisierte Spaltenfüllung, bei der zusätzlich neben dem Gang selbst auch das Hangende und Liegende bis zu einem gewissen Grade von Erz durchsetzt sind.

Sattelgang: Sattelförmiges Erzvorkommen, das zwischen oder in sedimentären Schichten im Bereich der scharfen Faltungsspitzen von Antiklinalstrukturen gebildet wird (Australien).

Tafelförmige Ganglagerstätte: Eine Serie nahe beieinander liegender, paralleler Bruchklüfte oder Spalten mit Erzfüllungen, die durch dünne Schichten bestehend aus Nebengestein getrennt sind. Nach einem Verdrängungsvorgang ergibt sich ein Gang, der alle Lagen erfaßt und durch eine Bänderstruktur gekennzeichnet ist.

Stufengang: Er verläuft wechselweise senkrecht (kreuzend) und parallel zur Schichtung des Nebengesteins.

Muldengang: Trogförmiges Erzvorkommen, das zwischen oder in sedimentären Schichten im Bereich der scharfen Faltungsmulden von Synklinalstrukturen gebildet wird (Australien).

Vein matter:
(vein material): The mineral content of an ore vein (see *vein*).

Material de veta: Contenido de mena en una veta (ver veta).

Caisse filonienne: Contenu minéral d'un filon métallifère (voir Filon).

Gangmaterial: Siehe Gang.

Vein system:
Framework or pattern of veins or of a simple vein; system of one set of parallel or a complex of mutually crosscutting, parallel or nearly parallel veins (see *vein*; network of veins).

Sistema de veta: Forma o modelo de vetas — sistema simple o complejo de una o más series de vetas paralelas o subparalelas que se cruzan (ver veta, y veta en malla.)

Système filonien: Ensemble de filons — Système simple ou complexe d'un groupe de filons parallèles ou de plusieurs groupes de filons parallèles ou approximativement parallèles se recoupant entre eux (voir filon, système filonien).

Gangsystem: Eine Vielzahl von Gängen — einfach oder komplex —; eine Serie von parallelen, teilweise parallelen oder sich wechselseitig durchkreuzenden Gängen (siehe auch Gang, Gangsysteme).

Volcanic exhalative mineral deposits (see Fig. 8, 9, 16; and App. VIIb):
Deposits formed from hydrothermal (ore) solutions exhaled from volcanic vents or fissures, principally onto the floor of the oceans or lakes.

Depósitos minerales exhalativos volcánicos: Depósitos formados desde soluciones (minerales) hidrotermales exhalados a través de respiraderos o fisuras volcánicas, principalmente sobre el fondo de los océanos o lagos.

Gîtes volcaniques exhalatifs: Gîtes formés à partir de solutions hydrothermales émises par des orifices volcaniques ou des fissures, surtout au fond de l'océan ou des lacs.

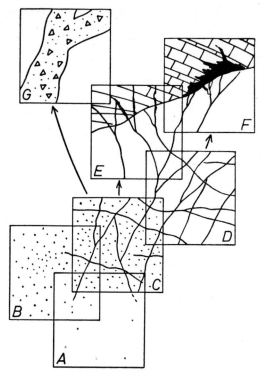

Fig. 16. Evolution of patterns in intrusive porphyry coppers. (G. C. Amstutz, 1963) (See also *ore*) (scale: 1 cm = 10 cm to 50 m).

Vulkanische Exhalationslagerstätten: Gebildet durch hydrothermale Lösungen, die in vulkanischen Schloten oder Spalten bis zum Boden von Ozeanen oder Seen aufsteigen.

Wall:

The surface between the country rock and the ore vein, ore layer or bed (hanging wall, foot wall).

Free wall: Relates to the wall of an ore vein in which the vein filling scales off cleanly from the gouge or wall rock.

Frozen wall: Relates to the boundary portion of a vein or layer which adheres tenaciously to the wall rock.

Pared: Superficie de la roca de caja que limita una veta o capa mineral bien definida (pared superior, pared inferior).

Pared libre: Se refiere a las paredes de una veta o capa mineral en cual el relleno de la veta se desprende limpio de la *salbanda* o de la roca de respaldo.

Pared adherida: Se refiere a la porción imítrofe del relleno de la veta o capa, la cual se adhiere tenazmente a las paredes de respaldo.

Mur: Surface de la roche encaissante délimitant un filon ou une couche avec précision (mur, toit).

Mur libre: Se rapporte au mur d'un filon on d'une couche dont le remplissage se sépare mécaniquement de la salbande ou de la roche encaissante.

Mur gelé: Se rapporte à la zone limitée d'un remplissage filonien ou d'une couche qui adhère fortement à la roche encaissante.

Wand: Die Fläche des Nebengesteins, die einen Erzgang oder ein Flöz begrenzt (Hangendes, Liegendes).

—: Hangendes oder Liegendes, von dem sich das Gang- oder Flözmaterial leicht ablöst.

—: Hangendes oder Liegendes, bei dem das Gang- oder Flözmaterial mit dem Nebengestein verwachsen („angebrannt") ist und eine klare Trennung Schwierigkeiten bereitet.

Wall rock:

The rock forming the walls of a vein, or lode, or bed, etc.; i. e., the country rock.

Pared de la caja: La roca que forma las paredes de una veta, filón, capa, etc., es decir la roca de caja.

Roche encaissante: Roche formant les murs d'un filon ou lode, ou d'une couche.

Nebengestein: Das unmittelbare Nachbargestein eines Erzganges, einer Erzmasse, oder eines Erzlagers.

Waste:

Valueless material such as barren rock to be thrown on the dump. Material too poor to mine or treat, as distinguished from ore. (McKinstry)

Desmonte: Material sin valor, tal como la roca estéril que se arroja a la cancha. Material muy pobre para ser trabajado o tratado, es decir, sin valor como mena.

Stérile: Matériau sans valeur, tel que les roches stériles, à déverser au terril. Matériau trop pauvre pour être exploité ou traité, par opposition au minerai.

Berge: Wertloses Material wie z. B. Abraum, das auf Halde geht. Auch Material mit zu geringen Gehalten, als daß es als Erz angesprochen und aufbereitet werden kann. Als Versatz verwendet nennt man es „Alter Mann".

Water level; Water table:

Surface, whether even or irregular, below which rock or soil is saturated with water. Above this surface, pores are incompletely filled and water is percolating downward. (McKinstry)

Nivel de agua; mesa de agua: Superficie, ya sea uniforme o irregular, debajo de la cual la roca o el suelo están saturados de agua. Encima de esta superficie los poros están parcialmente rellenados y el agua se filtra hacia abajo.

Niveau d'eau ou niveau phréatique: Surface, égale ou irrégulière, en dessous de laquelle la roche ou le sol sont saturés d'eau. Au-dessus de cette surface, les pores sont partiellement remplies et l'eau percole vers le bas.

Grundwasserstand: Die Oberfläche – glatt oder unregelmäßig – unterhalb der das Gestein oder der Boden mit Wasser saturiert ist. Oberhalb dieser Fläche sind die Poren nur unvollständig gefüllt oder wasserfrei und das Wasser sickert nach unten.

Whipstock:

A long, slender, steel wedge, with a groove along one side, that is placed at the bottom of the hole and is used to deflect the bit during drilling. (LEVORSEN)

Desviador: Cuña de acero larga y delgada con canales a un lado que se coloca al fondo de un sondaje y que sirve para desviar la corona durante la perforación.

Biseau de déflexion: Coin d'acier long et mince, présentant une rainure sur un côté, placé au fond d'un sondage et utilisé pour dévier l'outil au cours d'un forage.

—: Ein langer, schmaler Stahlkeil, mit einer Vertiefung in der abgeschrägten Seite, der in das Tiefste eines Bohrlochs eingeführt wird mit dem Zweck, die Richtung der weiteren Bohrung gewollt zu verändern.

Winze:

A downward opening like a shaft, but starting from a point underground rather than from the surface. It differs from a *raise* in that it is driven downward rather than upward. After a winze has connected with a level below it is described as a *raise* if viewed from the lower level. But in some mines the opening when connected through is called a *raise* no matter how driven originally or how viewed (McKINSTRY). In other mines such an opening is called a blind shaft.

Winze: Un hueco hacia abajo a manera de pique, pero que comienza en un punto subterráneo en vez de en la superficie. Se diferencia de una chimenea en que el winze se trabaja hacia abajo en vez de hacia arriba. Después de que un winze ha conectado con el nivel de abajo se llama chimenea, vista desde el nivel inferior. Pero en algunas minas, cuando el hueco alcanza a conectar se llama chimenea sin atender a cómo se trabajó o como se ve.

Descenderie: Ouverture dirigée vers le bas, pareille à un puits, mais creusée à partir d'un point de la mine plutôt qu'à partir de la surface. Elle diffère d'un «élevage» (*raise*) qui est plutôt creusé vers le haut. Une fois que la descenderie rejoint le niveau inférieur, elle est décrite comme montage, si elle est vue d'en bas. Dans certaines mines cependant, l'ouverture, après son achèvement, est désignée comme montage, peu importe le sens dans lequel on l'a creusée, ou dans lequel on la voit. Voir puits.

Gesenk: Ein Blindschacht, der von oben nach unten abgeteuft wird (Gegenteil: Aufbruch).

Xenothermal Ore Deposit:

Deposit formed at high temperature but at shallow to moderate depth. (McKINSTRY)

Depósito mineral xenotermal: Depósito formado a alta temperatura pero a poca o moderada profundidad.

Gîte xénothermal: Gîte formé à haute température, mais à profondeur faible ou modérée.

Xenothermale Lagerstätte: Gebildet unter hohen Temperaturen, jedoch in geringer Teufe.

Zone (see Figs. 5, 6a, 6b, 8, 9, 10a, 11, 13, 14, 17, 18, 19a, 19b; App. VIIb, c; VIIIc):

The designation *zone* may be given to any two or three dimensional area, and is a purely geometric term not implying any time or other genetic connotations. Zoning may be recognized on the basis of composition, or of fabric changes, or of both. A careful study may lead to a genetic understanding of the origin of zones. For example: Zone of weathering: Surficial portion of earth's crust with extensive oxidation and hydration (oxide and leaching zone).

Zone of cementation or enrichment: (Supergene sulphide zone) — intermediate area between oxide or leaching zone and primary ore zone in which a metal (e. g., Ag or Cu) accumulates by downward percolating solutions.

Zone of fracture: Relatively shallow environment where rock fails by fracture.

Zone of flowage: Deep portions of crust where rock fails by flowage (see also *biostratigraphic zone* and the definition of the term *zone* in the Code of Stratigraphic Nomenclature, Bull. A. A. P. G., vol. 45, No. 5, p. 645–665, 1961).

Zona: La denominación *zona* podría darse en cualesquiera de las dos o tres dimensiones de un área y es un término puramente geométrico que no entraña ninguna connotación genética de tiempo u otra cosa. Zonación podría reconocerse sobre la base de cambios de composición, fábrica, o ambos. Un cuidadoso estudio podría llevar a un conocimiento genético del origen de zonas; ejemplo:

Zona de intemperie: Porción superficial de la corteza terrestre, con extensiva oxidación e hidratación (zonas de lixiviación y oxidación).

Zona de cementación o enriquecimiento: (Zona de sulfuros supergénicos) — área intermedia entre la zona de oxidación o lixiviación y la zona de mineral primario, en la cual un metal (por ejemplo, Ag o Cu) se acumula por percolación descendente de soluciones.

Zona de fractura: Medio relativamente poco profundo, donde la roca se rompe por fractura.

Zona de deslizamiento: Porciones profundas de la corteza, donde las rocas se rompen por fluidez (ver también zona bioestratigráfica).

Zone: Le terme de zone peut être donné à n'importe quelle étendue bi-ou tridimensionelle. Il est purement géometrique et ne comporte aucune implication chronologique ou génétique. Un zonage peut être constaté dans les changements de composition ou de structure, ou dans les deux à la fois. Une étude détaillée peut conduire à une compréhension génétique de l'origine des zones. Exemples:

Zone d'altération: Partie superficielle de l'écorce terrestre soumise à une oxydation et á une hydratation étendues (zone d'oxydation et de lessivage).

Zone de cémentation ou d'enrichissement: Zone de sulfures supergènes. Partie intermédiaire entre la zone d'oxydation ou de lessivage et la zone de minerai primaire, dans laquelle un métal (p. ex. Ag ou Cu) s'accumule sous l'effet des solutions percollant vers le bas.

Zone de fracture: Environnement de faible profondeur où la roche cède par fracturation.

Zone de fluage: Parties profondes de l'écorce terrestre où la roche cède par fluage (Voir zone biostratigraphique).

Zone: Die Bezeichnung kann für jede zwei- oder dreidimensionale Einheit angewendet werden; sie ist eine rein geometrische Einheit, ohne Merkmale der Zeit oder der Genetik zu berücksichtigen. Zonen können bestimmt werden auf Grund von Kompositions-

114

merkmalen, oder von Veränderungen im Aufbau von Gesteinen. Sorgfältige Studien können zu einem genetischen Verständnis der Ursachen für Zonen führen. Z. B.:

Verwitterungszone: Oberflächennaher Teil der Erdkruste mit ausgedehnter Oxydation und Hydration (Oxydations- und Laugungszone).

Zementationszone: (*Supergene Sulphidzone*) Zone zwischen der Oxydations- und Auslaugungszone einerseits und der Primärzone andererseits, in der sich vorwiegend Metalle (z. B. Ag und Cu) durch abwärts sickernde Lösungen anreichern.

Bruchzone: Die unmittelbare Umgebung von Bruchstellen im Gestein.

Fließzone: Tiefe Krustenpartien („Unterbau"), wo das Gestein plastisch nachgibt. (Siehe auch „Biostratigraphische Zonen").

Zoning, zonal arrangement:

Arrangement of minerals or of fabric patterns in zones (no genetic connotation) (see also zone).

Zonación, disposicion zonal: El arreglo de minerales o de modelos de fábrica en zonas (sin connotación genética) (vease también zona).

Zonage: Disposition zonée de minéraux ou de structures. (Sans implications génétiques!) (voir aussi zone).

Zonare Verteilung: Anordnung von Mineralien oder von Struktursystemen in Zonen (kein genetischer Begriff) (vergleiche auch Zone).

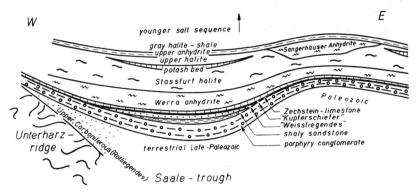

Fig. 17. Type pattern of syngenetic zoning in a depositional basin (schematic facies-profile in the Saale-trough, Mansfeld basin, Germany) (after STEINBRECHER).

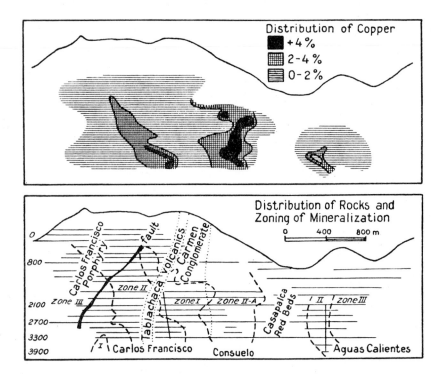

Fig. 18. Type pattern of epigenetic zoning by hydrothermal alteration and mineralization in a huge vein system cutting across all beds of the folded sediments and volcanics (Casapalca, Peru) (after K. OVERWEEL).

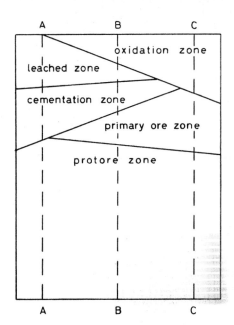

Fig. 19a. Scheme of supergene zoning in sulfide ore deposits.

The cuts *A*, *B*, *C* represent three basic possibilities produced by different climatic and groundwater conditions and erosion stages. The oxide zone may contain gossan, which is normally an iron oxide residue or accumulation, as well as rich oxide ores, the latter normally towards the bottom of this zone. The water table corresponds usually to the top of the cementation zone, which is often given the name of secondary sulfide zone, because of frequent chalcocite enrichments. The primary sulfide ore zone and the "protore" are often called hypogene zones. Protore is primary sulfide material below ore grade. (G. C. A.)

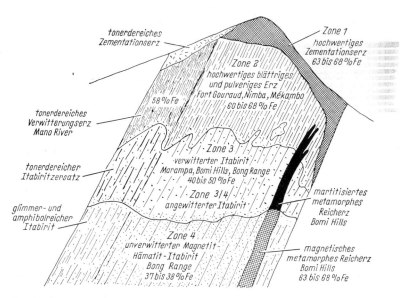

Fig. 19b. Example of supergene zoning in an iron deposit (after THIENHAUS, 1963), leading to enriched ore zones (translations of terms on next page).

tonerdereiches Zementationserz:
Alumina-rich supergene enrichment (cementation) ore; metal (mena) de enriquecimiento supérgeno (o de cementación) rico en alúmina; minerai de cémentation riche en aluminium.

tonerdereiches Verwitterungserz:
Alumina-rich "weathering ore" (oxide ore); metal (mena) de oxidación enriquecido en alúmina; minerai d'oxydation riche en aluminium.

tonerdereicher Itabiritzersatz:
Alumina-rich decomposed itabirite; itabirita decompuesta rico en alúmina; itabirite décomposée riche en aluminium.

glimmer- und amphibolreicher Itabirit:
itabirite rich in mica and amphiboles; itabirita con mucho mica y anfiboles; itabirite riche en mica et emphibole.

Zone 1, hochwertiges Zementationserz: zone 1, high grade supergene enrichment (or cementation) ore; zona 1, metal (mena) de alta ley de tipo de enriquecimiento; minerai de cémentation, très riche.

Zone 2, hochwertiges blättriges und pulveriges Erz; high grade, fissil or powdery ore; metal (mena) de alta ley, de textura en hojas o en polvo; minerai très riche feuilleté pulvérulent.

Zone 3, verwitterter Itabirit: weathered itabirite; itabirita decompuesta; itabirite altérée.

Zone 4, unverwitterter Magnetit-Hämatit-Itabirit: Unaltered magnetite-hematite-itabirite; itabirita no-alterada de magnetita y hematita; itabirite non-altérée à magnétite et hématite.

Martitisiertes metamorphes Reicherz: Rich metamorphic martitized ore; metal (mena) metamórphica martitizada; minerai métamorphique martitisé.

Magnetitisches metamorphes Reicherz: rich metamorphic and magnetitic ore; metal (mena) metamórphica y magnétitica; minerai riche métamorphique et magnétitique.

Appendices

Contents 119

(Note: The translations are given only where pertinent; separate loose sheets of some
of the tables are available from author.)

Appendix I

Measurements, weights and size scales

a) Units of measure (metric, incl. some conversions from non-metric systems)

Length		cm		*Area*		cm²
1 Ångström	(Å)	10^{-8}		1 Square millimeter	(mm²)	10^{-2}
1 Millimicron	($\mu\mu$)	10^{-7}		1 Square centimeter	(cm²)	1
1 Micron	(μ)	10^{-4}		1 Square decimeter	(dm²)	10^2
1 Millimeter	(mm)	10^{-1}		1 Square meter	(m²)	10^4
1 Centimeter	(cm)	1		1 Are	(a)	10^6
1 Decimeter	(dm)	10		1 Hectare	(ha)	10^8
1 Meter	(m)	10^2		1 Square kilometer	(km²)	10^{10}
1 Kilometer	(km)	10^5				

Volume (solid)		cm³		(liquid)		ml
1 Cubicmillimeter	(mm³)	10^{-3}		1 Milliliter	(ml) (cc)	1
1 Cubiccentimeter	(cm³) (cc)	1		1 Deciliter	(dl)	10^2
1 Cubicdecimeter	(dm³) (l)	10^3		1 Liter	(l)	10^3
1 Cubicmeter	(m³)	10^6		1 Hectoliter	(hl)	10^5

Weight		g
1 Milligram	(mg)	10^{-3}
1 Gram	(g)	1
1 Kilogram	(kg)	10^3
1 Ton	(t)	10^6

Length

1 Mile (statute) (mi) = 5280 ft. = 1.6094 km
(depth) 1 Fathom = 6 ft. = 72 in. = 1.8288 m
1 Yard (yd) = 3 ft. = 36 in. = 0,9144 m
1 Foot (ft) (′) = 12 in. = 30.48 cm
1 Inch (in) (″) = 2.54 cm

Area

1 Square mile (sq mi) = 640 acres = 2.5899 km²
1 Hectare (ha) = 2.47 acres = 10^4 m²
1 Acre (ac) = 0.4046 ha = 4840 square yards
1 Square yard (sq yd) = 9 square ft. = 0.8360 m²
1 Square foot (sq ft) = 144 sq in. = 929 cm²
1 Square inch (sq in) = 6.452 cm²

Volume

1 Cubic yard (cu yd) = 27 cu ft. = 0.7645 m³
1 Cubic foot (cu ft) = 1.728 cu in. = 0.02832 m³
1 Cubic inch (cu in) = 16.387 cm³

liquid

1 Barrel (crude petroleum) = 42 gallons
1 Gallon (gal) (US) = 4 quarts = 8 pints (pt) = 128 fluid ounces (fl oz) = 3.7852 l
1 Quart (qt) (US) = 0.9463 l
1 Gallon (gal) (Imperial) = 4 quarts = 8 pints = 160 fluid ounces = 4.546 l
1 Quart (qt) (Imperial) = 1.136 l

Weight

Avoirdupois

1 Long ton (lt) = 2240 pounds = 1.016 t
1 Short ton (st) = 2000 pounds = 0.907 t
1 Pound (lb av, or ♯) = 16 ounces (oz av) = 0.453 kg
1 Ounce (oz av) = 28.349 g

Troy

1 Pound (lb t) = 12 ounces = 0,373 kg
1 Ounce (oz t) = 31.103 g
1 Pennyweight (dwt) = $\dfrac{1}{20}$ oz

Pressures

	kg/cm²	lb/sq in	mm Hg	in Hg	water m/cm²	Bars
	1	14.22	735.6	28.96	10	0.9807
1 normal	1.0333	14.7	760	29.92	10.33	1.0132
atmosphere						
	0.070	1	51.71	2.036	0.7	0.0689
	0.0345	0.491	25.4	1	0.345	0.0338
	1.0197	14.504	759.9	29.53	10.19	1

Temperatures

	Centigrade °C (C 0/10)	Réaumur °R (R 0/10)	Kelvin °K (K-273/10)	Fahrenheit °F (F-32/18)
Freezing pt. H₂O	0	0	273	32
Boiling pt. H₂O	100	80	373	212

$$°F = \frac{9}{5} \; °C + 32$$

$$°K = °C + 273.15$$

$$°C = \frac{5}{9} \; (°F - 32)$$

b) Sizing, separation and observation scales

MESH (TYLER)		SEDIMENTS	OBSER-VATION	SIZING AND SEPARATING METHODS	EXAMPLES AND ELECTROMAGNETIC WAVE LENGTH IN Å
	256	COBBLE			
	128				
	64				
	32		naked eye + hand lens		(river gravel)
	10	PEBBLE — GRAVEL			$10^8 = 1 cm$
2½	8			screening	
5	4	granule		"sink and float"	(pea gravel)
9	2	very coarse		tabling	10^7
16	1	coarse		jigs	
		— SAND		F. = Flotation	
32	½	medium		"classification" = elutriation	(beach gravel)
60	¼	fine		sedimentation	10^6
115	⅛	very fine	"day light" microscopes		
250	1/16		infrared lights		(fine silt)
400	1/32	SILT			10^5
	1/64				(blood cells)
	1/128				
	1/256				(many germs)
	mm		visible light	centrifuge	10^4 - micron, μ
					(wave length visible ligth)
GAMMA RAYS		CLAY (WENTWORTH SCALE)	ultraviolet light	ultracentrifuge	10^3 (thinnest irides-cent films visible by light inter-ference)
			electron microscopes	colloids micro-organisms	10^2 (very large molecules)
			(hard) x-rays (soft)	clays and industrial slimes	
				cement rock material	$10 = 1$ millimicron, mμ
				taconite and certain sulfide ores	(average unit cell of crystals)
				most sulfide ores	
				phosphate, feldspar, and many industrial minerals	1 - Ångstrom, Å
				coal, potash, diamonds, etc	

Appendix II

Geochemical classifications of the elements

a) Periodic charts of the elements

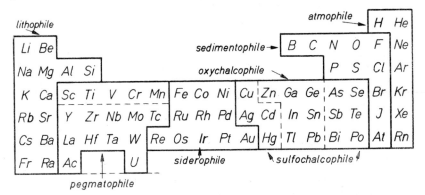

Geochemical subdivision of periodic chart of the elements, after SZADECZKY-KARDOSS (1955)

Geochemical subdivision of periodic chart of the elements, after VON PHILIPSBORN (1964)

b) *Classification of the elements according to their importance with respect to the formation of rocks and mineral desposits.*

1. *Chief elements of the lithosphere* O, Si, Al, H, Na, Ca, Fe, Mg, K, together with (especially from the atmosphere and biosphere) N and C (eleven elements). They are found chiefly in silicates and oxides, to a lesser extent in carbonates, sulphates phosphates, and nitrates.

2. *Important subsidiary elements of the lithosphere and hydrosphere.* They are very widespread, but, as a rule, in small quantities only. They sometimes replace other elements (Mn often in place of Al, Ca, or Fe; Ti often in place of Mg, Fe, or Al, rarely of Si).

3. *Incidental elements in mineral deposits of rock character* often occur in unusual mine rals but only locally in any quantity.

Light elements: Li, Be, B, F.

Rare-earths and related elements: Y, Ce, La, rare earths in general Zr, Th, U, Hf.

"Acid-Forming" elements in titanates, niobates, and tantalates: Nb, Ta.

Elements occurring in transitions to ore deposits: Mo, W, Sn (sometimes Bi), Cr, V.

4. *Incidental elements of mainly sulphophile character (also occurring in the native (metallic) state)*:

Triad elements: Ni, Co, Pt metals (beside the chief element, Fe).

Other noble metals: Au, Ag. They also occur combined with Se and Te.

In sulphides and sulphosalts (often with Fe): Cu.

Chiefly in sulphides, also in Ag sulphosalts: Zn, Pb, Hg.

Brittle metals (occur native, in special compounds, and in sulphosalts): As, Sb, and to some extent, Bi.

Some of these incidental elements show a tendency to replace more plentiful elements in widespread crystal compounds. They can then be included in the next group also.

5. *Camouflaged elements*:

The most typical trait of these elements is their disinclination to appear as the chief elements in any compounds. They tend, on the contrary, to appear in small quantities in widespread minerals, in which they partially replace some more important element. Hence they may be called diadochic elements. The fact that their presence in minerals is largely "camouflaged" has often led to underestimations of their abundance.

Rb, Cs, sometimes Tl (associated with K, sometimes with Ca).

Sr, Ba (associated with Ca, sometimes with K).

Sc (sometimes a substitute for Mg, Fe, Al, sometimes for Nb, Ta, rarely for W).

Ga (in the main a substitute for Al).

In (perhaps chiefly associated with Fe, Mn, and Sn).

Ge (chiefly associated with Si).

Hf (chiefly associated with Zr).

Re (chiefly associated with Mo).

Rarer metals of the Pd-Pt group (associated with Pt). (Similarly, Ni and Co are often associated with Fe).

Cd (associated with Zn).

Se and sometimes Te (associated with S).

In spite of its very summary nature, this characterization has an important bearing on mineral deposits. For it shows clearly why elements of similar average abundance may present very different problems in mining and extraction. By and large, the camouflaged elements are the most difficult to obtain on account of their very dispersed mode of occurence.

From: NIGGLI, P. (1954) Rocks and Mineral Deposits. W. H. Freeman & Co., San Francisco, p. 50–51.

c) Behaviour of the elements in oxides, etc.

occupied-space	coordination number rel. 0	1+	2+	3+	4+	5+	6+	7+
very small	usually III			B	C	N		
small	usually IV		Be		S	Cl Br	S	Cl
					Si Ge	P V As	Cr Se Mn Br	
medium	usually VI	Li Zn Mg Cu Pt-metals Fe Co Ni Cd Mn		Al Ga Fe V Cr Ti Mn Rh Sc In	Mn V Ti Mo Pt W Sn Nb Se Te Pb	Nb Ta Sb Bi	Mo W	J
relatively large	VI, > VI	Na K	Ca	rare earths	Zr Hf Ce U Th		IV IV	
large	usually > VI	Rb NH₄ Tl Cs	Sr Ba Pb					

The behaviour of the more important elements in oxides, hydroxides, salts of oxygenous acids etc. after P. NIGGLI (1949)

Appendix III

Geologic Column and Scale of Time

The table of this data sheet were prepared by Chester R. LONGWELL in cooperation with the Data Sheet Committee of the American Geological Institute. The radiogenic dates were selected and checked by A. KNOPF. (AGI data sheet 5).

Era	System and Period	Series and Epoch	Stage and Age North America	Stage and Age Europe	Absolute Age
Cenozoic	Quaternary	Recent			(Duration in years)[1] Approximately the last 10,000 years.
		Pleistocene	In glaciated regions (Glacial stages in italies)		
			Wisconsin[4]	*Würm*[5]	10,000 ± to > 35,000 years ago
			Sangamon	Würm-Riss	
			Illinoisan	*Riss*	
			Yarmouth	Riss-Mindel	
			.ansan	*Mindel*	
			Aftonian	Mindel-Günz	
			Nebraskan	*Günz*	
	Tertiary	Pliocene	(Atlantic and Gulf Coast)[6]	(Europe)	(Millions of years ago)
			Upper	Astian	
			Lower	Plaisancian	
		Miocene	Upper	Sahelian { Pontian / Sarmatian }	21[2]
			Middle	Tortonian / Helvetian	
			Lower	Burdigalian / Aquitanian	
		Oligocene	Upper / Middle / Lower	Chattian / Rupelian / Tongrian	
		Eocene	Jackson	Ludian / Bartonian	39[2]
			Claiborne	Auversian / Lutetian	
			Wilcox	Cuisian / Ypresian	
		Paleocene	Midway	Thanetian / Montian	60[3]

[1] Based on hundreds of radiocarbon dates.   [2] Determined by argon method;   [3] By strontium method.

126

Era	System and Period	Series and Epoch	Stage and Age		Absolute Age
			North America	Europe	
Mesozoic	Cretaceous	Upper (Late)	No accepted classification for North America generally.	Maestrichtian Campanian Santonian Coniacian Turonian Cenomanian	70^3 70^2 90^2
		Lower (Early)		Albian Aptian Barremian Hauterivian Valanginian Berriasian	
	Jurassic	Upper (Late)		Purbeckian Portlandian Kimeridgian Oxfordian	140^3
		Middle (Middle)		Callovian Bathonian Bajocian	
		Lower (Early)		Toarcian Pliensbachian Sinemurian Hettangian	
	Triassic	Upper (Late)		Rhaetian Norian Carnian	
		Middle (Middle)		Ladinian Anisian	
		Lower (Early)		Scythian	

These European names commonly used in North America also

March 1958

[4] Terminology for central North America;   [5] Terminology for Alps. Pleistocene marine stages also are recognized. [6] Other classifications are established for continental and Pacific Coast units.

Era	System and Period	Varied Subdivisions			Absolute Age (Millions of years ago)
Paleozoic	Permian	**Series and Epoch**	**Stage and Age**		220[1]
		(West Texas) Ochoa, Guadalupe, Leonard, Wolfcamp	Not established in North America	(Russia) Kazanian, Kungurian, Artinskian, Sakmarian	
	Carboniferous Systems — Pennsylvanian	(Central North America) Virgil, Missouri, Des Moines, Atoka, Morrow	Not established	(Europe) Stephanian, Westphalian, Upper Namurian	
	Carboniferous Systems — Mississippian	Chester, Meramec, Osage, Kinderhook	Not established	Lower Namurian, Viséan, Tournaisian	
	Devonian	**Series and Epoch**	**Stage and Age**		270[2]
		(Eastern United States) Bradfordian, Chautauquan	Conewango, Cassadaga, Chemung	(Europe) Famennian	
		Senecan	Finger Lakes, Taghanic	Frasnian	
		Erian	Tioughnioga, Cazenovia, Onesquethaw	Givetian, Eifelian	
		Ulsterian	Deerpark, Helderberg	Coblenzian, Gedinnian	
	Silurian	**Series and Epoch**	**Stage and Age**		
		(North American) Cayugan, Niagaran, Albion	(Britain) Downtonian, Ludlovian, Wenlockian, Valentian	Not established	
	Ordovician	**Series and Epoch**	**Stage and Age**	**Series and Epoch**	375
		(North America) Cincinnatian	Gamachian, Richmondian, Maysvillian, Edenian	(Britain) Ashgillian	
		Champlainian	Mohawkian, Chazyan	Caradocian, Llandeilian, Skiddavian	
		Canadian	Not established	Tremadocian	

Paleozoic	Cambrian	Upper (Late) Middle (Middle) Lower (Early)	Not established	Not established	440 470[3]
Era	System and Period	Varied Subdivisions			Absolute Age (Millions of years ago)
Precambrian	(No bases for worldwide divisions)	Latest Precambrian (?) (Wichita Mountains, Oklahoma)			550
		(Minnesota Scale) Keweenawan Group			1100[2]
		Pre-Keweenawan orogeny Animikie Group			1700[2]
		Algoman orogeny Knife Lake Group			2300[2]
		"Laurentian" orogeny Keewatin Group			2700[2]
		Oldest radiogenic date reported (Southern Rhodesia)			3310[2]

Chief sources of information on terminology in these tables are: Correlation charts and articles contributed by sections of Committee on Stratigraphy, National Research Council and published in Geol. Soc. Am. Bulletin: Cenozoic, vol. 54, p. 1713 fol.; Cretaceous, vol. 53, p. 435, fol., vol. 63, p. 1011 fol., vol. 65, p. 223 fol., vol. 55, p. 1005 fol.; Jurassic, vol. 63, p. 953 fol.; Triassic, vol. 68, p. 1451 fol.; Pennsylvanian, vol. 55, p. 657 fol.; Mississippian, vol. 59, p. 91 fol.; Devonian, vol. 53, p. 1729 fol.; Silurian, vol. 53, p. 533 fol.; Ordovician, vol. 65, p. 247 fol. Other sources: Mesozoic and Permian, M. Gignoux's "Stratigraphic Geology," 4th Ed., 1950 (English trans. by G. G. Woodford), Chaps. 6–8; also "Treatise on Invertebrate Paleontology, Part L, Mollusca," R. C. Moore, Ed., Geol. Soc. Am., 1957, p. 124–129; Outline of Precambrian, Minnesota Scale, from F. F. Grout et al., "Precambrian Stratigraphy of Minnesota" Geol. Soc. Am. Bull., vol. 62, 1951, p. 1017 fol.

[1] Determined by lead-uranium method; [2] By argon method; [3] By strontium method. Numbers underlined represent dates schecked by two or more methods. The radiogenic dates were selected and checked by A. KNOPF. [x] These provincial divisions are being extended widely in North America. [y] Divisions in Western Europe differ somewhat froms those in Russia.

Major parts of a (coal) mine

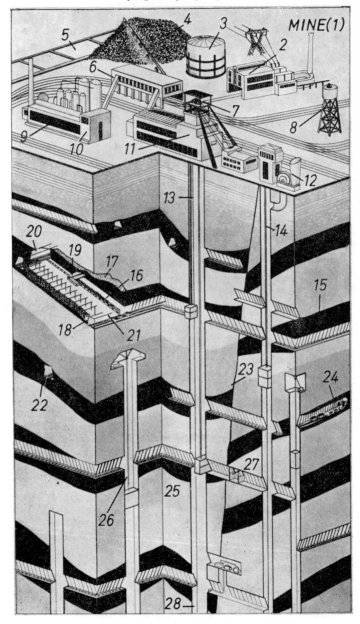

1 mine
mina
mine
Bergwerk

2 power station
planta eléctrica
centrale électrique
Kraftwerk

3 gasometer
gasómetro
gazomètre
Gasometer

4 dump
cancha (terrero, desechadero, escorial, montón, depósito)
terril
Halde

5 water clarifiers
tanques asentadores, estanques decantadores
bassins de décantation
Klärbecken

6 washery
lavadero (lavador, lavadora)
lavoir
Wäsche

7 hoisting tower
castillete
chevalement du puits
Fördergerüst

8 water reservoir
estanque, recipiente de agua (tanque, cisterna)
réservoir d'eau
Wasserspeicher

9 coking plant
planta de coque
cokerie
Kokerei

10 coal bunker
carbonera, tolva para carbón
silos de charbon
Kohlenbunker

11 hoisting house
sala de máquina de extracción
bâtiment d'extraction
Förderhaus

12 fan
ventilador
ventilateur
Ventilator (Lüfter)

13 main hoisting shaft
pique principal, pozo principal de extracción
puits principal
Hauptförderschacht

14 ventilation shaft
pozo de ventilación
puits d'aérage
Wetterschacht

15 coal seam, coal measures
capa de carbón
couche de charbon
Kohlenflöz

16 conveyer
transportador
convoyeur
Fördermittel

17 coal face
frente, frontón
front de taille
Strebfront

18 haulage road
galería de arrastre (o de transporte)
galerie de pied
Förderstrecke

19 channeling machine
rozadora
haveuse
Schrammaschine

20 return airway
galería de ventilación
galerie de tête
Wetterstrecke

21 train of mine cars
tren de vagonetas
train de berlines
Förderwagenzug

22 drift on the vein
galería de transporte (sobre capa)
voie en veine, voie de fond
Flözstrecke

23 fault
falla (dislocación)
faille
Verwerfung, Störung

24 mining with pneumatic picks
martillo picador (extracción por m. p.)
extraction au marteau pneumatique
Abbauhammergewinnung

25 cage
jaula
cage d'extraction
Förderkorb

26 blind pit with cage
pozo ciego o interior con jaula
puits intérieur (bure) avec cage d'extraction.
Blindschacht mit Förderkorb

27 ventilation door
puerta de ventilación
écluse d'aérage
Wetterschleuse

28 pit sump
pozo de desagüe
puisard
Schachtsumpf

Appendix V

Geologic and mine map symbols

Lithologic symbols for cross-sections and columnar sections; structural and mine-mapping symbols.
Símbolos litológicos para perfiles y columnas estratigráficas; signos para el mapeo estructural y minero.
Symboles lithologiques pour coupes et colonnes stratigraphiques; symboles employés en géologie structurale et en géologie minière.
Lithologische Zeichen für strukturelle und stratigraphische Profile; Zeichen für strukturelle und Minen-Kartierung.

a) Sedimentary rocks (sandy, argillaceous, calcareous).
Rocas sedimentarias (arenosas, arcillosas, calcáreas).
Roches sédimentaires (détritiques, argileuses, calcaires).
Sedimentäre Gesteine (sandig, tonig, kalkig).

1 a. massive sandstone (left)
arenisca maciza (izquierda)
grès massif (gauche)
massiger Sandstein (links)

b. bedded sandstone (right)
arenisca estratificada (derecha)
grès lité (droite)
lagiger Sandstein (rechts)

2 graded bedding
estratificación gradada
granoclassement positif
gradierte Schichtung

3 crossbedded sandstone
arenisca con estratificación cruzada
grès à stratification entrecroisée
kreuzgeschichteter Sandstein

4 breccia
brecha
brèche
Brekzie

5 conglomerate
conglomerado
conglomerat, poudingue
Konglomerat

6 calcareous sandstone
arenisca calcárea
grès calcaire
kalkiger Sandstein

7 shale partings in sandstone
intercalaciones arcillosas en arenisca
grès à passées argileuses
Tonlagen im Sandstein

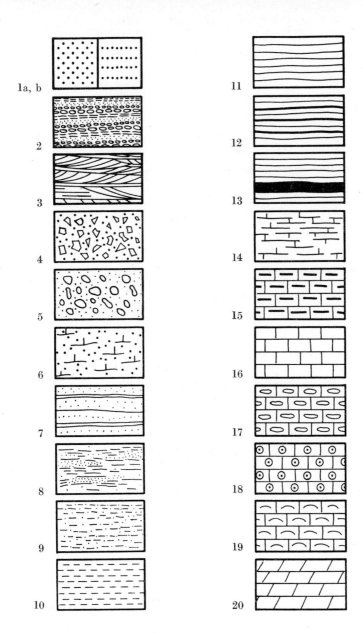

1a, b

2

3

4

5

6

7

8

9

10

11

12

13

14

15

16

17

18

19

20

8 sandstone lenses in shale
   lentes arenosas en lutita
   shale à lentilles de grès
   Sandsteinlinsen in Tonschiefer

9 siltstone
   "siltstone"
   aleurolite, microgrès
   Siltstein

10 mudstone, claystone, shale
   argilita
   argilite ou pélite
   Tonstein, Tonschiefer

11 shale, slate, marly slate
   lutita o pizarra
   shale, roche argileuse litée ou ardoise
   Mergelschiefer, Tonschiefer

12 oil shale
   esquisto petrolífero
   schiste bitumineux
   Ölschiefer

13 coal or ore in shale (or slate)
   capas de carbón o mineralización
   en esquistos
   niveau de charbon ou niveau minéralisé dans des schistes
   Kohle oder Vererzung im Tonschiefer

14 calcareous shale
   esquisto calcáreo
   shale calcaire
   Kalkschieferton

15 argillaceous limestone
   caliza arcillosa
   calcaire argileux
   toniger Kalk

16 limestone
   caliza
   calcaire
   Kalk

17 cherty limestone
   pedernal, caliza con nódulos silíceos
   calcaire à silex
   silexhaltiger Kalk

18 oölitic limestone
   caliza oolítica
   calcaire oolithique
   oolithischer Kalk

19 fossiliferous limestone
   caliza fosilífera
   calcaire fossilifère
   fossilreicher Kalk

20 dolomite
   dolomita
   dolomie
   Dolomit

b) *Volcanic and plutonic igneous and metamorphic rocks.*
Rocas ígneas volcánicas, plutónicas y rocas metamórficas.
Roches volcaniques, plutoniques et métamorphiques.
Eruptive und metamorphe Gesteine.

21 tuff – breccia
   brecha volcánica (tufácea)
   tuf bréchique
   Tuff – Brekzie

22 siliceous (left) and basic (right) lava
   flow
   lava ácida (izquierda) y básica (derecha)
   lave effusive acide (gauche), basique
   (droite)
   saure (links) und basische (rechts)
   Lava

23 granitic rock
   roca granítica

roche granitique
granitisches Gestein

24 Intermediate (left) and basic (right)
   igneous rock
   roca ígnea de composición intermedia
   (izquierda) o básica (derecha)
   roche plutonique intermédiaire (gauche) basique (droite)
   intermediäres (links) und basisches
   (rechts) Eruptivgestein.

25 schist (phyllite)
   esquisto (filita)
   schiste métamorphique
   Schiefer (Phyllit)

26 folded schist
   esquisto plegado
   schiste plissoté
   gefältelter Schiefer

27 gneiss
   gneis
   gneiss
   Gneis

28 marble
   mármol
   marbre
   Marmor

29 quartzite
   cuarcita
   quartzite
   Quarzit

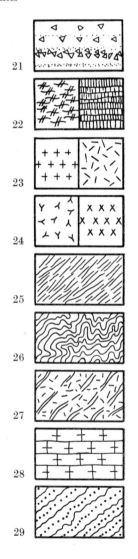

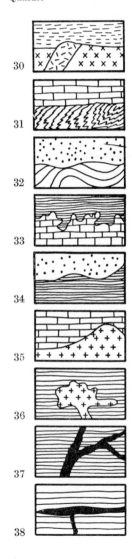

c) *Structural and geologic map symbols*

Structure symbols[1]
Símbolos estructurales
Représentation des structures
Struktur-Bezeichnungen

30 unconformity
discordancia estratigráfica
discordance
stratigraphische Diskordanz

31/32 nonconformity (or angular unconformity)
discordancia angular
discordance angulaire
Winkeldiskordanz

33 residual unconformity
discordancia residual
discordance sur karst
Rückstands-Diskordanz

---

[1] in part after SHROCK (1948)

Mapping symbols
Símbolos para mapas geológicos
Symboles de cartes géologiques.
Zeichen für geologische Kartierung

Contacts
Contactos
Contacts
Kontakte

39 contact, showing dip
contacto, con buzamiento
contact avec sens et valeur du pendage
Kontakt, mit Fallen

34 disconformity (or nonangular unconformity)
discordancia erosiva
lacune stratigraphique sous discordance angulaire sensible
Erosionsdiskordanz

35 buried hill
cúpula no aflorante
relief enterré (Paléorelief)
einsedimentierter Hügel

36 intrusion
intrusión
intrusion
Intrusion

37 dike
dique
filon ou dike
querschlägiger Gang

38 sill
capa intrusiva
sill
Lagergang

40 contact, vertical (left) and overturned (right)
contacto, vertical (izquerda) y volcado (derecha)
contact vertical (gauche) et renversé (droite)
Kontakt, senkrecht (links) und überkippt (rechts)

41 contact, approximate position
contacto, posición aproximada
contact placé approximativement
Kontakt, ungefähre Lage

42 Indefinite contact
contacto indefinido
contact hypothétique
unbestimmter Kontakt

43 concealed contact
contacto oculto
contact caché
verborgener Kontakt

44 gradational contact
contacto difuso o gradacional
contact progressif
Kontakt mit unbestimmtem Übergang

39 ―――― 20   42 ▪▪▪▪▪▪▪▪▪▪▪

40 ―|― 90/30   43 ··········

41 ― ― ―   44 ― ― ― ―

Bedding, foliation, joints, veins and dikes
Estratificación, foliación, diaclasas, vetas y diques
Stratification, foliation, diaclases, filon et dikes
Schichtung, Schieferung, Klüfte, Adern und Gänge

45 bedding, dip
 estratificación, buzamiento
 stratification, pendage
 Schichtung, Fallen
46 overturned bedding, dip
 estratificación volcada
 stratification renversée
 Überkippte Lage, Schichtung
47 bedding (top shown by primary features)
 estratificación (techo indicado por
 elementos primarios)
 stratification dont le toit est donné
 par des figures de sédimentation
 primaires
 Schichtung, (Oberseite durch primäre
 Elemente belegt)
48 horizontal bedding
 estratificación horizontal
 stratification horizontale
 waagrechte Schichtung

49 vertical bedding (point indicates
 stratigraphic top)
 estratificación vertical (punto-techo
 estratigráfico)
 stratification verticale (le point indiquant le toit)
 vertikale Schichtung (Punkt zeigt
 stratigraphisch Hangendes an)
50 crumpled beds, general dip
 capas onduladas, buzamiento general
 couches plissotées, pendage général
 gefältelte Schichtung, generelles
 Fallen
51 strike and dip uncertain
 rumbo y buzamiento indefinido
 direction et pendage mal définis
 Streichen und Fallen unsicher
52 dip uncertain
 buzamiento mal definido
 pendage mal défini
 Fallen unsicher

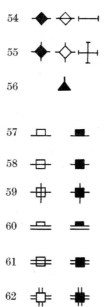

53 foliation
foliación
foliation
Schieferung

54 vertical foliation
foliación vertical
foliation verticale
senkrechte Schieferung

55 horizontal foliation
foliación horizontal
foliation horizontale
horizontale Schieferung

56 foliation parallel to bedding
foliación paralela a la estratificación
foliation parallèle à la stratification
Schieferung parallel der Schichtung

57 joints
diaclasas
diaclases
Klüfte

58 vertical joints
diaclasas verticales
diaclases verticales
senkrechte Klüfte

59 horizontal joints
diaclasas horizontales
diaclases horizontales
horizontale Klüfte

60 veins and dikes
vetas y diques
filons et dikes
Adern und Gänge

61 vertical veins and dikes
vetas y diques verticales
filons et dikes verticaux
senkrechte Adern und Gänge

62 horizontal veins and dikes
vetas y diques horizontales
filons et dikes horizontaux
horizontale Adern und Gänge

Folds
Pliegues (plegamientos)
Plis
Falten

Anticlines (left)   Synclines (right)
Anticlinales (izquierda)   Sinclinales (derecha)
Anticlinaux (gauche)   Synclinaux (droite)
Antiklinalen (links)   Synklinalen (rechts)

63 trace of axial plane and plunge of axis
traza del plano axial e inclinación del eje
trace du plan axial (avec indication du plongement de l'axe)
Spur der Achsenebene und Eintauchen der Achse

64 asymmetric (steeper limb = double arrow)
asimétrico (ala próxima a la vertical = doble flecha)
asymétrique (la double flèche indiquant le flanc de plus incliné)
asymmetrisch (steilerer Schenkel mit Doppelpfeil)

65 overturned (trend and plunge of axis)
volcado (dirección e inclinación del eje)
renversé (avec indication de la direction et du plongement de l'axe)
überkippt (Richtung und Eintauchen der Achse)

66 doubly plunging
domo
double terminaison périclinale
zweiseitig geneigt

67 vertically plunging
con eje vertical
plongement vertical de l'axe
senkrecht eintauchend

68 inverted limb
ala invertida
flanc retourné
überkippter Mittelschenkel

Faults
Fallas
Failles
Verwerfungen

69 fault, showing dip
falla, con buzamiento
faille avec sens et valeur de pendage
Verwerfung (Fallen)

70 fault, vertical
falla vertical
faille verticale
senkrechte Verwerfung

71 fault, approximate
falla no bien definida
faille tracée approximativement
Verwerfung, ungefährer Verlauf

72 fault, uncertain
falla supuesta
faille hypothétique
Verwerfung, Verlauf unsicher

73 fault, concealed
falla oculta
faille cachée
Verwerfung verdeckt

74 fault, possible (as from aerial photo-
graphs)
falla posible (de la interpretación
de aerofotos)
faille possible (déduite p. ex. de la
photo aérienne)
Verwerfung wahrscheinlich (aus Luft-
photointerpret.)

75 fault, (U or circle = upthrown side;
D or dot = downthrown side), trend
and plunge of linear features
falla (U o círculo = lado elevado, D
o punto = lado hundido), dirección
e inclinación de señas lineares

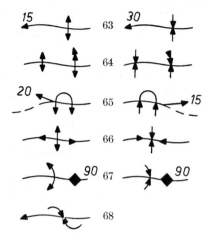

faille (U ou o = lèvre relevée (up),
D ou • = lèvre abaissée (down), direction approximative et plongement
de figures linéaires
Verwerfung (U oder Kreis = geho-
bene Seite, D oder Punkt = versenkte
Seite), Richtung und Eintauchen
linearer Elemente

76 relative lateral movement
movimiento horizontal relativo
mouvement latéral relatif
relative Horizontalverschiebung

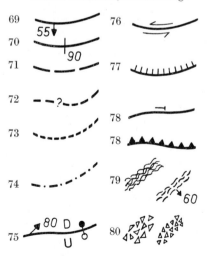

139

77 normal fault (hachures on down-thrown side)
fálla normal (sombreado sobre lado hundido)
faille normale (les hachures sont du côté de la lèvre abaissée)
normale Verwerfung, Abschiebung (Striche an versenkter Seite)

78 thrust or reverse fault (T or sawteeth on upper plate)
falla inversa (T o aserrado sobre lado elevado)
faille inversé (le T ou les triangles sont du côté de la lèvre relevée)
inverse Verwerfung, Aufschiebung oder Überschiebung (T oder Zähne an überschobener Platte)

79 fault or shear-zones (showing average dip)
zonas de fallamiento o cizallamiento (mostrando inclinación aproximada)
zone de faille ou de cisaillement, sens et valeur approximative du pendage
Verschiebungs- oder Abscherungszonen (mit durchschnittlichem Fallen)

80 fault breccia
brecha de falla
brèche de faille
Verwerfungsbrekzie

d) *Mine map symbols*

Signos para el mapeo minero
Symboles pour mines et sondages
Zeichen zur Gruben-Kartierung
Surface
Superficie
Au jour
Oberfläche

81 vertical shaft
pique vertical
puits vertical
senkrechter Schacht

82 inclined shaft
pique inclinado
puits incliné
Schrägschacht

83 open adit
socavón

81
82
83
84
85
86
87
88
89

Underground:

90
91
92
93
94
95
96

97
98
99
100
101
102
103
104
105
106
107
108
109
110
111
112

galerie accessible
Stollen, Mundloch

84 inaccessible adit
socavón inaccesible
galerie inaccesible
unzugänglicher Stollen

85 trench
trinchera
tranchée
Schürfgraben

86 prospect
prospección
prospect
Schürfung

87 mine or quarry (working; abandoned)
mina o cantera (activa; abandonada)
mine ou carrière (gauche en activité;
droite abandonnée)
Grube oder Steinbruch (in Betrieb;
aufgelassen)

88 open pit or quarry
tajo abierto o cantera
exploitation à ciel ouvert ou carrière
Tagebau oder Steinbruch

89 sand or gravel pit
cantera de arena o ripio
sablière ou ballastière
Sand- oder Kiesgrube

Underground workings
Subterráneo
Au fond
Untergrund

90 raise or winze (head; foot)
coladero o pozo ciego (techo; base)
montage ou descenderie (gauche-tête)
(droite-pied)
Aufbruch oder Blindschacht (Gesenk)

91 entry of tunnel
boca de socavón
entrée de galerie
Stolleneingang

92 adit with open cut
socavón con tajo abierto
galerie précédée d'une tranchée
Stollen mit Schürfung

93 level working (ore chute, above; inaccessible, right)
galería de labor (chimenea para
mineral, arriba; inaccesible, derecha)
galerie de niveau (gauche: chute)
(droite: inaccessible)
Abbausohle (Erzrolle, oben; unzu-
gänglich, rechts)

94 logging or cribbing along drift
enmaderado a lo largo de una galería
garnissage le long d'une galerie
Holzeinbau längs Strecke

95 filled workings
labor rellenada
travaux remblayés
versetzter Abbau

96 drill hole, inclined (projected end of
hole)
perforación dirigida (proyección del
fondo de la perforación)
sondage incliné (projection)
Bohrung, gerichtet (Ende projektiert)

97 oil well
pozo petrolífero
puits de pétrole
Ölbohrung

98 gas well
perforación para gas
puits de gaz
Gasbohrung

99 dry well
perforación seca
puits sec
trockene Bohrung

100 well with showings of oil
perforación con manifestaciones de
petróleo
puits avec indices de pétrole
Bohrung mit Ölvorkommen

101 well with showings of gas
perforación con manifestaciones de
gas
puits avec indices de gaz
Bohrung mit Gasvorkommen

102 abandoned oil well
pozo de petróleo abandonado
puits de pétrole abandonné
verlassene Ölbohrung

141

103 abandoned gas well
   pozo de gas abandonado
   puits de gaz abandonné
   verlassene Gasbohrung

104 shut oil well
   perforación de petróleo tapada
   puits de pétrole obturé
   verschlossene Ölbohrung

105 shut gas well
   perforación para gas tapada
   puits de gaz obturé
   verschlossene Gasbohrung

106 water well
   perforación para agua
   puits d'eau
   Grundwasserbohrung

107 nonflowing water well
   perforación para agua no surgente
   puits sans écoulement naturel
   nicht fließende Grundwasserbohrung

108 dry water well
   perforación para agua seca
   puits sec
   trockene Grundwasserbohrung

109 ore vein (trace or mapped shape; dip)
   veta mineralizada (traza generalizada
   o forma actual; buzamiento)
   filon de minerai (forme et pendage)
   Erzader (Spur oder kartierte Form;
   Fallen)

110 high grade ore (with disseminated ore
   in altered wallrock)
   mineralización de alta ley (con
   mineral diseminado en roca de caja
   alterada)
   minerai riche, entouré de minerai
   disséminé (roche encaissante altéré)
   hochgrädiges Erz (mit feinverteiltem
   Erz im umgewandelten Nebengestein)

111 low grade orebody
   cuerpo mineralizado de baja ley
   minerai pauvre
   erzarmer Körper

112 mill or mine dump
   relaves o cancha de mina
   terril de mine ou de laverie
   Berge, Halde

Appendix VI

Fabric and locking charts

A Geometric Classification of Basic Intergrowth Patterns of Minerals

A connotation-free set of purely descriptive patterns, 1) for studies of rocks and mineral deposits, particularly for the present revision of genetic theories, 2) for ore dressing microscopy, metallography, and other fields of applied petrology, mineralogy, and metallurgy.

Between most of these nine common locking types there are naturally gradational transitions with regard to both pattern and size. Particle or grain size data are a prerequisite of any accurate study of rocks and mineral deposits and enhance the value of this chart.

a) Mineral intergrowth or locking chart

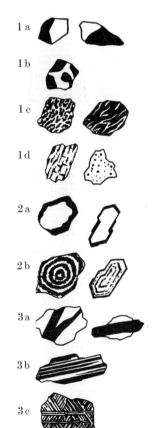

Type 1a Simple intergrowth or locking type; rectilinear or gently curved boundaries. Most common type, many examples.

Type 1b Mottled, spotty, or amoeba-type locking or intergrowth. Simple, common pattern; many examples.

Type 1c Graphic, myrmekitic, or "eutectic" type. Common; examples; chalcopyrite and stannite; quartz and feldspars; etc.

Type 1d Disseminated, emulsion-like, drop-like, buckshot or peppered type. Common; examples: chalcopyrite in sphalerite or stannite; sericite, etc. in feldspars; tetrahedrite in galena; etc.

Type 2a Coated, mantled, enveloped, coronarim-, ring-, shell-, or atoll-like. Common; examples; chalcocite or covellite around pyrite, sphalerite, galena; etc.; kelyphite rim, and other rims.

Type 2b Concentric-spherulitic, or multiple shell-type. Fairly common; ex.: uraninite with galena, chalcopyrite, bornite; cerussite-limonite; Mn- and Fe-oxides; etc.

Type 3a Vein-like, stringer-like, or sandwich-type. Common; ex.: molybdenite-pyrite; silicates; carbonates; phosphates; etc.

Type 3b Lamellae-, layered, or polysynthetic type. Less common; examples: pyrrhotite-pentlandite; chlorite-clays; etc.

Type 3c Network, boxwork, or Widmanstaetten-type. Less common; ex.: hematite-ilmenite-magnetite; bornite or cubanite in chalcopyrite; millerite-linnaeite; metals, etc.

G. C. Amstutz — 1954, 1960

143

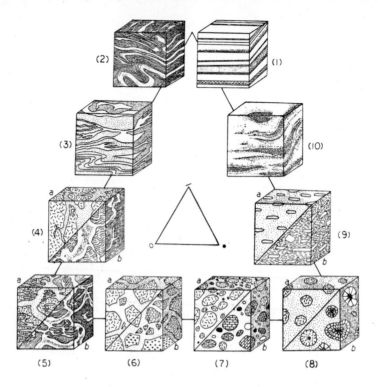

b) Large scale rock and ore fabric chart

Geometric Classification of Rock Fabrics

free of genetic connotations

Pattern 1, 2, 3, and 10 = stromatitic or linear fabrics
Pattern 4, 5, and 6   = merismitic or network fabrics
Pattern 7, 8, and 9   = ophthalmitic or disseminated fabrics
Pattern 11   = massive or homogeneous fabric (to be placed at the fourth
corner of a tetrahedron of which this drawing shows one
face only).

(This purely geometric classification and nomenclature of rock fabrics is a systematic modification of patterns pictured by PAUL NIGGLI in "Rocks and Mineral Deposits," Basel 1948, San Francisco 1954, for chorismatic, polyschematic rocks and mineral deposits. Additional adjectives may be used in order to designate transitional patterns: pattern 3 may be called phlebitic stromatite, pattern 4a phlebitic merismite, pattern 8b miarolithic ophthalmite, and pattern 10 nebulitic stromatite.)

G. C. AMSTUTZ, 1959,
(Proc. Geol. Assoc. Canada 11, p. 104)
(and AGI data sheet 21)

Appendix VII

a) *Classification of major ore deposits*

(After RAMDOHR, 1960, with minor modifications)

Metamorphic sequence	Metamorphism under stress	32 Katazone 31 Mesozone 30 Epizone	30
		29 Contact-metamorphism	
Sedimentary Sequence	Ore deposition through precipitation	28 Mineral fuels (Caustobiolites) 27 Sedimentary sulfide deposits 26 Concentration in arid basins 25 Descendent dikes and replacements	25
	Ore deposition through weathering	24 Cementation-zone (often enriched) 23 Oxydation-zone 22 Placer deposits	
Magmatic sequence	Volcanic-(extrusive-) magmatic	21 Mixed deposits (partic. "sterile hot springs") 20 Volcanic-exhalative deposits 19 Volcanic-intramagmatic	20
	Intrusive magmatic — Subvolcanic hydro-thermal	18 Hg-Sb-Formation 17 Cu-Pb-Zn-Formation 16 Au-Ag-Formation	
	Hydro-thermal stage	15 Sulfide-lean-formation 14 Cu-siderite-formation 13 Ag-Sn-Zn-formation 12 Ag-Co-Ni-U-Bi-As-Formation 11 Low-temperature Zn-Pb-Formation 10 Pb-Zn-Ag-Formation 9 Cu-As-Fe-Formation 8 Au-Fe-Formation	15 10
	Pegmatit. pneumato-lytic stage	7 Pneumatolytic impregnations 6 Contact-pneumatolytic deposits 5 Pneumatolytic dikes 4 Pegmatites	5
	Intra-magmatic stage	3 Intrusive ore-injections 2 Liquidmagmatic unmixing 1 Crystallisation differentiates	

b) *Subdivisions and nomenclature of magmatic ore deposits* (partly after SCHNEIDERHOEHN, 1955)

plutonic ore deposits

genetic groups of ore deposits	approximate temperature	classification according to distance from parent magma
hydrothermal — epithermal	100°	5) telemagmatic
mesothermal	250°	4) kryptomagmatic
katathermal	400°	3) apomagmatic
pneumatolytic – pegmatitic	600°	2) perimagmatic
liquid magmatic (melt, parent magma)		1) intramagmatic

depth zones of ore deposits in regard to the parent batholith

volcanic and subvolcanic ore deposits

lavas and exhalative mineral deposits

subvolcanic ore deposits

subvolcanic rocks

batholithic rocks

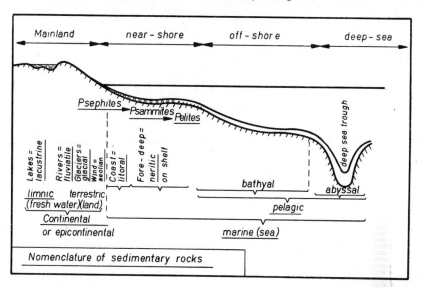

Nomenclature of sedimentary rocks

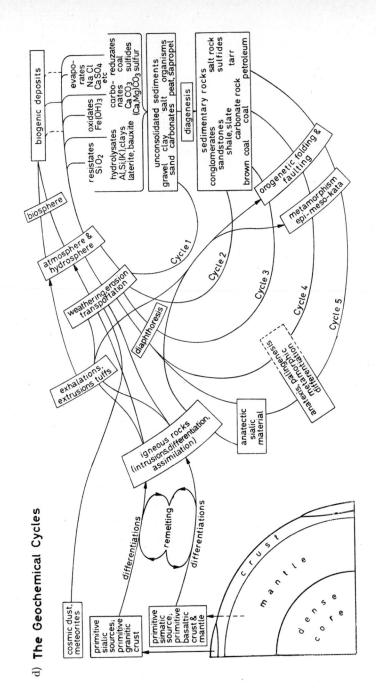

d) **The Geochemical Cycles**

Appendix VIII

Crystallization and paragenesis of magmatic rocks and ores

a) *The "reaction series"*

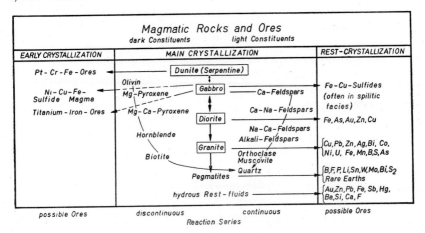

Reaction Series of Crystallization Differentiation
after ROSENBUSCH, NIGGLI, BOWEN, RAMDOHR, SCHÜLLER, with modifications

b) *Crystallization diagrams*

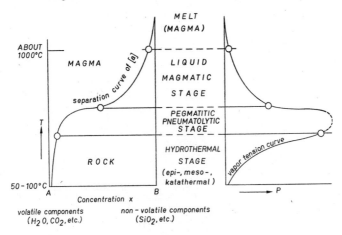

Crystallization diagrams for magmas, after P. NIGGLI (1929)

c) *Paragenetic chart of magmatic ore minerals*

	Liquid-magmatic	Pegmatitic	Pneumatolytic		Hydrothermal		Formations	
			Cassiterite Pneumat.	Tour-Qtz (Au) Veins	Me²·Fe Sulfides–Au	Cu–Pb–Zn	Bi–Co–Ni	Sb–Hg
Pt	native Pt, Sperrylite							
Cr	Chromite, Cr-spinel							
Ni	Pentlandite, Pyrrhotite, rare Ni-arsenides						Ni-Arsenides etc	
Fe	Pyrrhotite, Pyrite, Magnetite, Hematite etc							
Mo				Molybdenite		Wulfenite		
W		Wolframite		Scheelite				
Ti	Perovskite, Titanite, Ilmenite, Ti-Silicates, Titanates, Rutile							
Nb,Ta		Tantalates, Niobates with Titanates						
Zr,Th	Zircon	Zr-Silicates	Monazite					
Li		Li-Silicates-Phosphates						
Be		mostly Beryl						
Sn			Cassiterite		Stannite			
Ge			Ge-Sulfosalts					
Co	in Sulfides			Cobaltite			Co-As-Ores	
V								
Cu		Chalcopyrite		Chalcopyrite, As-Sb Sulfosalts				
Zn				Sphalerite				
Pb				Galena		Sulfosalts		
Bi(U)	native Bi + Bismuthinite		Uraninite				native Bi	
Au			native Au, Sulfides, Tellurides					
Ag			Me*-Fe Sulfides, Argentite, As-Sb-Sulfosalts,			native Ag		
As			Arsenopyrite		Cu-Sulfosalts	Tennantite		Realgar
Sb					Tetrahedrite			Stibnite
Te			mostly with Gold					
Se			mostly with Gold, Silver					
Ba					Barite	Witherite		
Sr					Strontianite, Celestite			
Hg	scantly in Sulfosalts							Cinnabar
CO₂		Rare-earth carbonates			various carbonates		Calcite	
P	Apatite	various Phosphates						
Cl	Sodalite,	Scapolite, Apatite, enters in Scapolite						
B	Tourmaline	Boron-Silicates, Borates		Tourmaline				
S								

* Me = Ag, As, Cu, Ni, Co, Sn (Ni, Co, As compounds also without Fe and S)

Paragenesis chart of magmatic ore minerals, after P. NIGGLI (1929), with modifications

Appendix IX

Hardness scale

Old MOHS'scale	Hardness classes	Microhardness kg/mm^2	Hardness classes	New MOHS'scale
Diamond – 10				Industrial diamond (111)
	15	10000	15	Diamond – Carbonado
	14	9000 — 8000	14	
	13	7000 — 6000	13	$B_{6,5}C$
	12	5000	12	B_4C
	11	4500 — 4000	11	Boron
	10	3500 — 3000	10	TiC
Corundum – 9	9	2500	9	Corundum(11$\bar{2}$0)
Topaz – 8	8	2000 — 1500	8	Topaz(001)
Quartz – 7	7	1000	7	Quartz(10$\bar{1}$1)
Orthoclase – 6	6	900 800 700	6	Magnetite(111)
Apatite – 5		600 500 400		Scheelite(111)
	5	300	5	
Fluorite – 4	4	200	4	Fluorite(111)
Calcite – 3	3	100	3	Galena(100)
Gypsum – 2	2	50	2	Halite(100)
Talc – 1	1	10 5	1	Talc(001)
		1		

MOHS' hardness scale of minerals, after POWARRENNYCH (1963)

Appendix X.

Metal-content of the most important ore-minerals, arranged according to the Periodic System.
(in part after SCHNEIDERHÖHN, 1962).

Atom. Nr.	Element	Mineral	Formula	Metal-content in % theoretical	Metal-content in % actual	Density	Hardness (MOHS)
3	Lithium Li	Spodumene	$LiAl[Si_2O_6]$	3.73 Li	1.34–3.43 Li	3.1–3.2	6½–7
		Amblygonite	$LiAl[F, OH/PO_4]$	4.7 Li	3.3–4.67 Li	3	6
		Triphylite / Lithiophilite	$Li(Fe, Mn)[PO_4]$	4.4 Li		3.5	4½–5
		Lepidolite	$KLi_2Al[(F, OH)_2/Si_4O_{10}]$ (?) $KLi_{1.5}Al_{1.5}[(F, OH, ^1/_2O)_2/AlSi_3O_{10}]$	3.56–3.69 Li to 2.61–2.76 Li	0.6–2.76 Li	2.8–2.9	2½–4
		Zinnwaldite	$KLiFeAl[(F, OH)_2/AlSi_3O_{10}]$	1.58–1.60 Li		2.9–3.1	
		Petalite	$Li[AlSi_4O_{10}]$	2.2 Li		2.4	6½
11	Sodium Na	Halite	$NaCl$	39.3 Na		2.2	2½
19	Potassium K	Sylvite	KCl	52.4 K		2	2
		Carnallite	$KCl \cdot MgCl_2 \cdot 6 H_2O$	14.1 K		1.6	1
		Leucite	$K[AlSi_2O_6]$	18 K		2.45	5½–6
		Kainite	$K, Mg[Cl/SO_4] \cdot 3 H_2O$	15.7 K		2.1	3
37	Rubidium Rb	Carnallite	see above		0.015–0.037 Rb		1
		Lepidolite	see above		1.19–3.46 Rb		2½–4
53	Cesium Cs	Lepidolite	see above		0.075–0.68 Cs		2½–4
		Pollucite	$(Cs, Na) [AlSi_2O_6] \cdot H_2O_{<1}$	42.8 Cs		2.9	6½
4	Beryllium Be	Beryl	$Al_2Be_3 [Si_6O_{18}]$	5.07 Be		2.7	7½–8
		Chrysoberyl	Al_2BeO_4	7.15 Be		3.7	8½
		Helvite	$(Mn, Fe, Zn)_8[S_2/(BeSiO_4)_6]$		2.8–5.4 Be	3.1–3.4	6
12	Magnesium Mg	Magnesite	$MgCO_3$	28.8 Mg		3.0	3½–5
		Dolomite	$CaMg[CO_3]_2$	13.15 Mg		2.9	3½–4
		Carnallite	see above	8.74 Mg		1.6	1
		Kieserite	$MgSO_4 \cdot H_2O$	17.6 Mg		2.57	3½
		Olivine	$(Mg, Fe)_2 [SiO_4]$	34.4 Mg		3.3	6½–7

Atom. Nr.	Element		Mineral	Formula	Metal-content in % theoretical	Metal-content in % actual	Density	Hardness (Mohs)
20	Calcium	Ca	Calcite	$CaCO_3$	40 Ca		2.7	3
			Anhydrite	$CaSO_4$	29.4 Ca		3.0	3–3½
			Gypsum	$CaSO_4 \cdot 2\,H_2O$	23.3 Ca		2.3	2
38	Strontium	Sr	Strontianite	$SrCO_3$	59.3 Sr		3.7	3½–4
			Celestite	$SrSO_4$	47.7 Sr		3.9	3½
56	Barium	Ba	Barite	$BaSO_4$	58 Ba		4.3	3–3½
			Witherite	$BaCO_3$	69.5 Ba		4.5	3½
13	Aluminum	Al	Hydrargillite (Gibbsite)	$Al(OH)_3$	34.7 Al		2.3	3
			Diaspore	$AlOOH$	45.0 Al		3.4	6½–7
			Cryolite	$Na_3[AlF_6]$	12.8 Al		2.9	2½
			Kaolinite	$Al_4[(OH)_8/Si_4O_{10}]$	20.9 Al		2.6	2–2½
			[Nepheline]	$KNa_3[AlSiO_4]_4$				5½–6
39	Yttrium	Y	Gadolinite	$Y_2FeBe_2[O/SiO_4]_2$	38 Y		4.5	6½–7
			Samarskite	$(Y, Er)_4[(Nb, Ta)_2O_7]_3$	20–28.4 Y	1.6–11.7 Y	6	5–6
57 71	Cerium etc.	Ce	Monazite	$Ce[PO_4]$	59.7 Ce	17.1–30.3 Ce	5	5–5½
			Orthite (Allanite)	$(Ca, Ce, La, Na)_2)Al, Fe, Be, Mg, Mn)_3[O/OH/SiO_4/Si_2O_7]$	5.5 Ce		3–4	5½–6
			Pyrochlor (Koppite)	$(Na, Ce, Fe, Ca)_2(Nb, Ta, Ti)_2 O_6(OH, F, O)$		1.14–15.4 Ce 5.88–6.89 Ce	4.4–4.5	5–5½
			Cerite	$(Ca, Fe)Ce_3H[(OH)_2/SiO_4/Si_2O_7]$ (? with La, Dy, Al)		20–54 Ce	4.9	5½
14	Silicon	Si	Quartz	SiO_2	46.7 Si		2.6	7
22	Titanium	Ti	Rutile	TiO_2	60.0 Ti		4.2	6–6½
			Ilmenite	$FeTiO_3$	31.6 Ti		4.5	5½–6

Atom. Nr.	Element		Mineral	Formula	Metal-content in %		Density	Hardness (MOHS)
					theoretical	actual		
40	Zirconium	Zr	Zircon Baddeleyite	Zr[SiO$_4$] ZrO$_2$	49.7 Zr	up to 70 Zr	4.5 4.9–5.4	7½ 6½
90	Thorium	Th	Thorite Monazite	Th[SiO$_4$] see above	71.7 Th	2.02–24.1 Th	4.6 5	5 5–5½
23	Vanadium	V	Patronite Descloizite Mottramite Vanadinite Carnotite Roscoelite	VS$_4$ Pb(Zn,Cu)[VO$_4$/OH] Pb$_5$[(VO$_4$)$_3$/Cl] (K,Na,Ca,Cu,Pb)$_2$ [(UO$_2$)/(VO$_4$)$_2$]$_2$·3 H$_2$O V-bearing muscovite	11.6 V 10.8 V	28–39 V 9.8–13.7 V 11.3–12.8 V 4.38–16.1 V	6 6.7–7.2 7 2.9–3	± 2 3½ 3 no report 2½
41	Niobium	Nb	Columbite Pyrochlor (Koppite) Fergusonite	(Fe, Mn) (Nb, Ta)$_2$O$_6$ see above Y(Nb,Ta)O$_4$	51.4 Nb	22–54.5 Nb 43–48 Nb 20–32 Nb	5.3–7.3 4.4–4.5 4.3–6.2	6 5–5½ 5½–6
73	Tantalum	Ta	Tantalite Fergusonite	(Fe,Mn) (Ta,Nb)$_2$O$_6$ see above		43–66 Ta 1.6–22 Ta	6.5–8.2 4.3–6.2	6 5½–6
24	Chromium	Cr	Chromite	FeCr$_2$O$_4$	46.4 Cr		4.8	5½
42	Molybdenum	Mo	Molybdenite Wulfenite Powellite	MoS$_2$ Pb[MoO$_4$] Ca[MoO$_4$]	60 Mo 26.1 Mo 48 Mo		4.8 7.0 4.4–4.5	1–1½ 3 3½
74	Tungsten (Wolfram)	W	Ferberite Hübnerite Scheelite	Fe[WO$_4$] Mn[WO$_4$] Ca[WO$_4$]	60.5 W 60.7 W 63.8 W	Wolframite	7.5 7.1 6.0	5 5 4½–5

Atom. Nr.	Element	Mineral	Formula	Metal-content in %		Density	Hardness (Mohs)
				theoretical	actual		
92	Uranium U	Uraninite (Pitchblende)	$(U, Th)O_2$	83.3 U	up to 76.7 U	9.5	5–6
88	Radium Ra	Carnotite	see above		up to 55 U	7	
25	Manganese Mn	Pyrolusite	MnO_2	63.2 Mn		up to 5	1–2
		Psilomelane					5–6
		Cryptomelane					6–6½
		Polianite					
		Manganite	$Mn_2O_3 \cdot H_2O$	62.46 Mn		4.3	4
		Braunite	$Mn^{2+}Mn_6^{4+}[O_8/SiO_4]$	63.6 Mn		4.8	6–6½
		Hausmannite	Mn_3O_4	72.0 Mn		4.7	5½
		Rhodochrosite	$MnCO_3$	47.8 Mn		3.5	3½–4½
		Rhodonite	$Mn[SiO_3]$	41.9 Mn		3.5	5½–6
		Jacobsite	$MnFe_2O_4$	23.8 Mn		4.7	5½
26	Iron Fe	Hematite	Fe_2O_3	70.0 Fe		5.2	5½–6½
		Magnetite	Fe_3O_4	72.35 Fe		5.2	6
		Limonite	$FeOOH$	62.85 Fe		up to 4	5–5½
		Siderite	$FeCO_3$	48.21 Fe		3.0–3.8	3½–4
		Chamosite	$(Fe, Fe^{3+})_3 [(OH)_2/AlSi_3O_{10}]$ $(Fe, Mg)_3(O, OH)_6$		28.5–37.3 Fe	3–3.4	no indic.
		Ilmenite	$FeTiO_3$	36.8 Fe		4.5	5½–6
27	Cobalt Co	Skutterudite and Smaltite	$CoAs_{3-2}$	28.23 Co		6.5	4½–5
		Safflorite	$CoAs_2$			6.2	5½
		Cobaltite	$CoAsS$	35.52 Co		4.8–5.8	4½–5½
		Linnaeite etc.	$(Co, Ni)_3S_4$		11–53 Co	2–4	5–6
		Asbolane	Co-bearing psilomelane		3.15–27 Co	2–4	4½
		Heterogenite (Stainierite)	$Co(OH)_2$	63.6 Co			

Atom. Nr.	Element		Mineral	Formula	Metal-content in %		Density	Hardness (Mohs)
					theoretical	actual		
28	Nickel	Ni	Niccolite	NiAs	43.92 Ni		8	5–5½
			Chloantite and Rammelsbergite	NiAs$_2$	28.14 Ni		6.2–7.2	5+
			Pentlandite	(Fe, Ni)$_9$S$_8$		10–40 Ni	4.5–5	3½–4
			Garnierite	(Ni,Mg)$_6$[(OH)$_8$ Si$_4$O$_{10}$]		4.3–36.1 Ni	2.3–2.8	2–3
	Platinum Metals	Pt	Ferroplatinum	Pt, Fe		75–84 Pt	14–19	4–4½
44	Ruthenium	Ru				2–4 Ir		
45	Rhodium	Rh						
46	Palladium	Pd	Platiniridium	Pt, Ir		20–67 Pt	20–22	7
76	Osmium	Os				19–69 Ir		
77	Iridium	Ir	Newjanskite	Ir, Pt, Os, Rh, Ru		1–12 Pt		
78	Platinum	Pt	Sysserskite	Os, Ir, Ru, Pt, Rh		44–70 Ir 17–40 Os 2–12 Rh 7–9 Ru 0–12 Pt 17–30 Ir 38–68 Os 0–4 Rh 9–14 Ru 51 Ir 26 Os 19 Au 3 Ru	19–21 up to 21	6–7
			Aurosmiridium	Ir, Os, Au, Ru				

Atom. Nr.	Element	Mineral	Formula	Metal-content in % theoretical	Metal-content in % actual	Density	Hardness (Mohs)
78	Platinum (Pt)	Osmite	Os, Ir, Rh		80 Os, 10 Ir, 5 Rh		
		Sperrylite	$PtAs_2$	56.6 Pt		10	6–7
		Cooperite	PtS	86.5 Pt		9	4
		Stibiopalladinite	Pd_3Sb	70 Pd		0.5	4–5
29	Copper (Cu)	native Copper	Cu	100 Cu		9	2½–3
		Chalcocite	Cu_2S	79.9 Cu		5–6	2½–3
		Covellite	CuS	66.5 Cu		4.6	1½–2
		Chalcopyrite	$CuFeS_2$	34.7 Cu		4.2	3½–4
		Bornite	Cu_9FeS_6 to Cu_3FeS_3	63.3 Cu to 55.0 Cu		4.9–5.5	3
		Enargite	Cu_3AsS_4	48.4 Cu		4.4–4.5	3
		Tetrahedrite	Cu_3SbS_{3-4}		23–45 Cu	4.4–5.4	3–4½
		Tennantite	Cu_3AsS_{3-4}		30–53 Cu	5.8	2½–3
		Bournonite	$CuPbSbS_3$	13.0 Cu		6	3½–4
	(Only oxidation zone)	Cuprite, Tenorite	Cu_2O, CuO	88.8 Cu, 57.5 Cu		4	6
		Malachite	$Cu_2[(OH)_2/CO_3]$	55.3 Cu		3.8	3½–4
		Azurite	$Cu_3[(OH)/CO_3]_2$	40.4 Cu		3.3	5
		Dioptase	$CuSiO_3 \cdot H_2O$	59.0 Cu		3.7	3–3½
		Atacamite	$Cu_2(OH)_3Cl$	25.5 Cu		2.2	2½
		Chalcantite	$CuSO_4 \cdot 5\,H_2O$	56.2 Cu		3.9	4
		Brochantite	$Cu_4[(OH)_6/SO_4]$				
47	Silver (Ag)	native Silver	Ag	100 Ag		10	2–3
		Acanthite, Argentite	Ag_2S	87 Ag		7	2–2½
		Proustite	Ag_3AsS_3	65.4 Ag		5.6	2–2½

Atom. Nr.	Element	Mineral	Formula	Metal-content in % theoretical	Metal-content in % actual	Density	Hardness (Mohs)
47	Silver[1]	Pyrargyrite	Ag_3SbS_3	59.8 Ag		5.8	2½
		Stephanite	Ag_5SbS_4	68.3 Ag		6.2	2–2½
		Polybasite	$Ag_{16}Sb_2S_{11}$	75.5 Ag		6	2–3
		Pearceite	$Ag_{16}As_2S_{11}$	78.4 Ag		6.1	
	Only oxydation zone	Cerargyrite	$AgCl$	75.0 Ag		5.5–5.6	1–1½
79	Gold[2] Au	native Gold	Au, Ag		80–98 Au	15–19	2½–3
		Electrum	Ag, Au		70–75 Au	12–16	2½–3
		Calaverite	$AuTe_2$	43.7 Au		9	1½–2
		Sylvanite	$AuAgTe_4$	24.2 Au		8	1–1½
		Nagyagite	$AuTe_2 \cdot 6\,Pb(S,\,Te)$		6–13 Au	6.8–7.5	
		Petzite	$(Ag,\,Au)_2Te$		up to 75 Au	8.7–9	2½–3
30	Zinc Zn	Sphalerite	$(Zn,\,Fe,\,Mn,\,Cd)S$ pure ZnS		up to 67 Zn	3.9–4.1	3½–4
		Wurtzite			67.1 Zn	5	4
		Franklinite	$(Zn,\,Mn)(Fe,\,Mn)_2O_4$		7–20.5 Zn	5	6
		Willemite	Zn_2SiO_4	58.6 Zn		4	5½
		Zincite	ZnO	80.3 Zn		5.5	4–4½
	Only oxidation zone	Smithsonite	$ZnCO_3$	52.1 Zn		4.3	5
		Hemimorphite	$H_2Zn_2SiO_5$	54.3 Zn		3.5	4½–5
48	Cadmium Cd	Sphalerite	$(Zn,\,Fe,\,Mn,\,Cd)S$		0.05–3.2 Cd	3.9–4.1	3½–4
		Smithsonite	$(Zn,\,Cd)CO_3$		0.02–0.8 Cd	4.3–4.4	5
80	Mercury Hg	Cinnabar	HgS	86.2 Hg	up to 17 Hg	8	2½
		Schwazite (Hg-tetrahedrite)	$(Cu_2,\,Hg)_3Sb_2S_6$			5	3½–4½

[1] The main silver production is from: galena, tetrahedrite chalcocite, pyrite whose Ag-concentration varies widely.
[2] Other main gold-ores are pyrite, arsenopyrite, stibnite etc.

Atom. Nr.	Element		Mineral	Formula	Metal-content in % theoretical	Metal-content in % actual	Density	Hardness (Mohs)
31	Gallium	Ga	Germanite	$Cu_3(Fe,Ge,Ga)S_4$		up to o,8 Ga	4.3	3
			Gallite	$CuGaS_2$		?	4.4	3½
			Sphalerite	see above		traces	4.0	3½-4
49	Indium	In	Sphalerite	see above		traces		
81	Thallium	Tl	Pyrite	see below		traces		
32	Germanium	Ge	Germanite	$Cu_3(Fe,Ge,Ga)S_4$		up to 8 Ge	4.3	3
			Renierite	$Cu_3(Fe,Ge)S_4$, Fe-rich Germ.			± 4.3	– 4½
50	Tin	Sn	Cassiterite	SnO_2	78.7 Sn		6.8	6-7
			Stannite	Cu_2FeSnS_4	27.6 Sn		4.4	4
			Cylindrite	$Pb_3Sn_4Sb_2S_{14}$ (?)	24.8 Sn		5.4	2½-3
			Teallite	$PbSnS_2$	30 Sn		6.4	1½
82	Lead	Pb	Galena	PbS	86.6 Pb		7.5	2½
			Bournonite	$CuPbSbS_3$	42.4 Pb		5.8	2½-3
			Boulangerite	$Pb_5Sb_4S_{11}$	55.2 Pb		6.1	3½-3
	Only oxidation zone {		Cerussite	$PbCO_3$	77.55 Pb		6.5	3-3½
			Anglesite	$Pb[SO_4]$	68.33 Pb		6.3	3
			Pyromorphite	$Pb[Cl/(PO_4)_3]$	76.38 Pb		7.0	3½-4
33	Arsenic	As	Arsenopyrite	$FeAsS$	46.0 As		6	5½-6
			Loellingite	$FeAs_2$	72.8 As		7	5-5½
51	Antimony oxidation zone	Sb	Stibnite	Sb_2S_3	71.7 Sb	75 Sb	4.5	2
			Antimonyochre	$(Ca,NaH)Sb_2O_6(O,OH,F)$			5.1-5.2	variable
83	Bismuth	Bi	native Bismuth	Bi	100 Bi		9.8	2-2½
			Bismuthinite	Bi_2S_3	81.3 Bi		6.5	2

Atom. Nr.	Element		Mineral	Formula	Metal-content in %		Density	Hardness (Mohs)
					theoretical	actual		
16	Sulfur	S	native Sulfur	S	100 S		2	$1\frac{1}{2}-2\frac{1}{2}$
			Pyrite	FeS_2	53.4 S		5.2	$6-6\frac{1}{2}$
			Pyrrhotite	$Fe_{1-x}S$	36.5 S		4.6	4
			Gypsum	$Ca[SO_4] \cdot 2H_2O$	23.2 S		2.3	2
34	Selenium	Se	Pyrite	see above		traces		
			Sphalerite	see above		traces		
			Selenides					
52	Tellurium	Te	Calaverite	$(Au, Ag)Te_2$	56.4 Te		9.0−9.4	$2\frac{1}{2}$
			Sylvanite	$AuAgTe_4$	62.6 Te		7.9−8.3	$1\frac{1}{2}-2$
			Nagyagite	$AuTe_2 \cdot 6Pb(S, Te)$ (?)		18−30 Te	6.8−7.5	$1-1\frac{1}{2}$
			Petzite	$(Ag, Au)_2Te$	24.4 Te		8.7−9.0	$2\frac{1}{2}-3$
			Hessite	Ag_2Te	37.1 Te		8.2−8.9	$2\frac{1}{2}-3$

Bibliography

of selected dictionaries, glossaries, lexica, etc. (some of which have been consulted or quoted with the permission of the publishers; note abbreviations in parentheses, e. g. AGI, McKINSTRY, SCH.).

A. G. I. Glossary of Geology and Related Sciences (1960) American Geological Institute, Washington, 325 + 72 p. (AGI)

DENNIS, J. G. (1967) International Tectonic Dictionary. AAPG, Tulsa, Oklahoma, 196 p.

HUEBNER, W. (1939) Geology and Allied Sciences. Veritas Press. New York, 2 Vol., both 405 p.

JONES, W. & A. CISSARZ (1931) Englisch-deutsche geologisch-mineralogische Terminologie. Murby, London; Weg, Leipzig und Nostrand, New York, 250 p.

McKINSTRY, H. E. (1949) Mining Geology. Prentice-Hall, New York, 680 p.

MURAWSKI, H. (1963) Geologisches Wörterbuch. Enke, Stuttgart, 243 p.

NOVITZKY, A. (1951) Dictionario Minero-Metalurgico-Geologico-Mineralogico-Petrografico y de Petroleo. Buenos Aires, 369 p.

NOVITZKY, A. (1958) Indice Alfabetico para el Dictionario Minero (1951). Buenos Aires, 217 p.

SCHIEFERDECKER, A. A. G. (1959) Geologische Nomenklatur. Noorduijn, Gorinchem, Holland, 533 p. (SCH)

Index

The use of the INDEX is straight forward and simple: The four languages are separated. All the terms which are defined or described are in bold face letters, including the pages on which these definitions or descriptions are given. APPENDIX numbers and pages are given as follows: V (138), the last number being the page on which the term is listed. For space reasons this first edition of the Glossary does not give translations for all parts of the APPENDIX. The author and the editor would be very glad to receive additions and corrections.

El uso del INDEX es sencillo: Los cuatro idiomas estan separados. Todos los términos definidos o descritos estan escritos con letras gruesas como tambièn las paginas respectivas. Referencias al APPENDIX son dadas como sigue: V (138), el segundo numero es la pagina. En esta primera edicion, no se dan todas las traducciones del APPENDIX. El autor ruega información sobre los errores.

L'usage de l'INDEX est simple: les quatre langues sont séparées. Tous les termes définis ou décrits sont mis en lettres grasses, comme aussi les pages ou cette information plus complète est donnée. Les termes dans l'APPENDIX sont indiqués de la manière suivante: V (138), c'est à dire, la page est entre parenthèses. Pour cette première édition pas tous les textes de l'APPENDIX sont traduits. L'auteur et l'éditeur seront très contents de recevoir des suggestions et des corrections.

Der Gebrauch des INDEX ist einfach: Alle definierten oder beschriebenen Begriffe sind fettgedruckt, sowie auch die betreffenden Seitenzahlen. Auf Begriffe im APPEN-DIX wird wie folgt hingewiesen: V (138), wobei die Zahl 138 die Seite angibt. Aus Platzgründen sind einige Übersetzungen im APPENDIX ausgelassen. Die Herausgeber bitten um Ergänzungsvorschläge und Korrekturen.

Disconformity (symbol), V (135, **136**)
Dolomite (symbol), V (133, **134**)
Doubly plunging fold (symbol), V
    (**138**, 139)
Drag 22
    ore 22, **65**
Drift 22
    (glacial) 22
Drift on the vein, IV (131)
Drill hole (symbol), V (140, **141**)
Drilling 8, 9, 13, 21, 23, 29, 87
    , cable tool 9
    , churn 9, 13
    , diamond 13
    , directional 21
    mud 23
Drill bit 8
    pipe 23
Druse 24, 36
Dry well (symbol), V (140, **141**)
Dump, IV (131)
    , mill or mine (symbol), V (140, **142**)

Echelon 24, 25
Edelfall 24
Edle Geschicke 24
Elements,
    , geochemical classifications II (123
        to 124); metallic & most important
        ore-minerals X (152—160)
    , behaviour in oxides, II c (125)
    , camouflaged II b (124)
    , hydroxides, salts, etc., II c (125)
    , periodic chart (Table), II a (123)
    , sulphophile, II b (124)
Eluvial placers 25, Fig. 14 (80)
Endogenetic 25, 27
Endogenic 25, 27
Endogenous 25, 27, 43
En echelon, en échelon 24, 25
Enrichment, zone of 114
Environments, sedimentary, VII c (147)
Eolian placers 25
Epigenetic 25
    zoning, Fig. 18 (116)
Epimetamorphism 2
Epithermal 26, 103
Exhalation 26
Exhalative deposits 27, Fig. 8 (67) **110**
Exogenetic 25, **27**, 100

Exogenic 25, 27
Exogenous 25, 27
Exsolution minerals 27

Fabric 27
    , mineral (intergrowth), VI a (143)
    , rock and mineral deposits, VI b (144)
Face (mining) **28**
Face, coal, IV (131)
Facing (of strata), **28**
Fahlband 28
Fan, IV (131)
Fault 28
Faults (symbols), V (139—140)
Fill, waste 29
Filled workings (symbol), V (140, **141**)
Filter pressing 29
Fishing 29
Fissure 30
    systems, Fig. 2 a (30)
    patterns, Fig. 2 b (31)
    vein 31, **106**
Flat (see pitch) **31**
Float 32
Flotation 32
Flow, siliceous and basic lava (symbol),
    V (**134**, 135)
Flowage, zone of 114
Flux 32
Fold 33
Fold patterns, Fig. 3 a, b (33)
    trace of axial plane and plunge of
        axis (symbol), V (**138**, 139)
Folded schist (symbol), V (135)
Folds (symbol), V (138—139)
Foliation (symbol), V (137, **138**)
Formation 34
Formulas, ore minerals, X (152—160)
Fossiliferous limestone (symbol), V (133,
    134)
Fractional crystallization 34
Fracture 28, **34**
    cleavage 28, 34
    system 35
    , zone of 114
Free wall 35, **111**
Frozen wall 35, **111**

Gangue 35
Gas well (symbol), V (141)

168

184

190